本书的出版得到北京外国语大学"双一流"建设重大标志性项目经费资助

数字技术与外语教育丛书

大语言模型的外语教学与研究应用

Applications of Large Language Models in Foreign Language Teaching and Research

许家金　赵　冲　孙铭辰　|编著|

外语教学与研究出版社
FOREIGN LANGUAGE TEACHING AND RESEARCH PRESS
北京 BEIJING

图书在版编目（CIP）数据

大语言模型的外语教学与研究应用 = Applications of Large Language Models in Foreign Language Teaching and Research：汉文、英文 / 许家金，赵冲，孙铭辰编著 . -- 北京：外语教学与研究出版社，2024.2（2024.4 重印）
（数字技术与外语教育丛书）
ISBN 978-7-5213-5048-7

I. ①大… II. ①许… ②赵… ③孙… III. ①自然语言处理 - 应用 - 外语教学 - 教学研究 - 汉、英 IV. ①TP391②H09

中国国家版本馆 CIP 数据核字 (2024) 第 037502 号

出 版 人	王　芳
选题策划	段长城
项目负责	都帮森
责任编辑	步　忱
责任校对	李晓雨
装帧设计	梧桐影
出版发行	外语教学与研究出版社
社　　址	北京市西三环北路 19 号（100089）
网　　址	https://www.fltrp.com
印　　刷	北京盛通印刷股份有限公司
开　　本	650×980　1/16
印　　张	15.5
版　　次	2024 年 2 月第 1 版 2024 年 4 月第 3 次印刷
书　　号	ISBN 978-7-5213-5048-7
定　　价	69.90 元

如有图书采购需求，图书内容或印刷装订等问题，侵权、盗版书籍等线索，请拨打以下电话或关注官方服务号：
客服电话：400 898 7008
官方服务号：微信搜索并关注公众号"外研社官方服务号"
外研社购书网址：https://fltrp.tmall.com

物料号：350480001

前　言

不论承认与否，人们的工作和生活都将因大语言模型而发生深刻变化。一些原本重复、机械的工作，在大语言模型的辅助下效率倍增；一些原本复杂、困难的操作在大语言模型时代成为可能。将大语言模型与工作、生活结合，人们得以从烦琐的流程中解脱出来，赢得时间，从事更富价值、更具创意的事情；人们也有可能因此放慢节奏、增进交往、贴近自然。

大语言模型算法的技术飞跃，给外语圈带来的是惊异和焦虑。生成式人工智能的性能跃升，引发人们对语言教学和语言学的下行担忧。然而，作为技术局外人，在技术层面我们难以入局、变局。与其让忧虑占据思绪，不如凑近上前，一探究竟。新的人工智能替代的或许是本该被替代的苦役，只是这一进程比预想的来得更早一些。先行体验大语言模型的语言工作者，已经深感其在翻译、文字润色等方面的高效和便利。本书旨在探索将大语言模型系统应用于外语教学与研究的种种可能，从而推动外语教学与研究的提质升级。本书是北京外国语大学语料库语言学研究团队在这方面的初步尝试，希望能对学界同仁善用智能技术、改进教学与课程、推动学科发展有所助益。

在人工智能技术高速发展的今天，有关大语言模型的书籍，写成之时，便可能是过时之日；出版之际，已有再版之需。技术进步之快，无法预期。因此，我们特设配套网页（https://corpus.bfsu.edu.cn/info/1082/1914.htm），将书中涉及的提示语和大语言模型反馈结果上传，同时会将后续技术进步及时同步至该配套网页。若遇网址变动，读者可通过搜索引擎查询《大语言模型的外语教学与研究应用》配套网页"访问该页面。

书中所用提示语以英文给出，主要考虑到国际主流大语言模型的主体训练语料为英文。在实践中，我们也发现在大语言模型中使用以英文写成的提示语，返回的结果更为理想。读者可根据个人习惯采用汉语或其他语言自行尝试。总体上，提示语的语种不会造成大语言模型生成结果出现显著差异。

在本书的酝酿阶段，北外语料库语言学研究团队全体成员在课程、沙龙、聚餐、游览时，时时切磋、处处启发，使得我们对大语言模型在外语教学与研究中的应用场景和运用思路方面的思考不断深化、愈发系统。我们深切感到，"大语言模型语言学研究方法"呼之欲出。虽有不完善、不确定之处，但本书所尝试的分析思路大体上已涉及语言学研究的方方面面。更重要的是，很多原本完全依赖专家思辨、人工解读的分析任务，如今大语言模型的表现也不输人类，转瞬完成，真正让外语学科更上层楼。

本书第一作者要感谢在书稿写作过程中家人的巨大付出，这保证了书稿的顺利完成。同时，也感谢两位合作作者提供了大量有价值的案例与思路。另外，本书的三位作者都有一个先天不足，即我们对语言教学缺乏深入了解。因此本书所举案例在教学价值和合理性方面可能有所欠缺，还请读者斧正。希望我们的案例能在方法和技术实现层面为广大教师提供启示。在外语研究相关案例的选择方面，限于我们的研究偏好，难免在视野上以偏概全。欢迎学界同仁为我们提供新的语言研究应用场景，共同推进大语言模型的语言学应用。

本书系教育部人文社会科学重点研究基地重大项目"基于多语种语料库的外语及外语教育研究"（22JJD740012）的阶段性成果。

许家金、赵冲、孙铭辰
北京外国语大学
中国外语与教育研究中心 / 人工智能与人类语言重点实验室

目录

前言 ·· i

第一章　大语言模型概述

1.1 大语言模型的定义和特点 ·· 2

1.2 大语言模型在外语教学与研究中的应用概述 ···································· 3

1.3 外语教学与研究中的提示工程方法 ·· 6

第二章　大语言模型在外语教学中的应用

2.1 大语言模型在词汇教学中的应用 ·· 10

　　2.1.1　学习包含词根 port 的英语单词 ··· 10

　　2.1.2　围绕特定主题造句成段语言促成设计 ···································· 11

　　2.1.3　指定英语词汇的读后续写材料设计 ·· 14

　　2.1.4　fare 和 fee 的词义辨析 ··· 16

　　2.1.5　cloth、cloths、clothing 和 clothes 的词义辨析 ······················ 17

　　2.1.6　suggest 和 advise 的用法辨析 ·· 19

　　2.1.7　get around 的熟词生义问题 ·· 21

　　2.1.8　拼写练习的设计 ··· 23

2.2 大语言模型在语法教学中的应用 ·· 25

　　2.2.1　读后续写关系从句学习材料设计 ··· 25

　　2.2.2　读后续写虚拟语气学习材料改编 ··· 27

iii

		2.2.3	宾语补语的用法 ⋯⋯⋯⋯⋯⋯⋯⋯⋯⋯⋯⋯⋯⋯⋯⋯⋯⋯⋯⋯	30
		2.2.4	过去完成体和过去完成进行体的区分 ⋯⋯⋯⋯⋯⋯⋯⋯⋯	32
		2.2.5	否定副词引导的部分倒装句学习 ⋯⋯⋯⋯⋯⋯⋯⋯⋯⋯⋯	34
		2.2.6	主谓一致问题 ⋯⋯⋯⋯⋯⋯⋯⋯⋯⋯⋯⋯⋯⋯⋯⋯⋯⋯⋯⋯	37

2.3 大语言模型在听力教学中的应用 ⋯⋯⋯⋯⋯⋯⋯⋯⋯⋯⋯⋯⋯⋯ 40

    2.3.1 听力材料改编 ⋯⋯⋯⋯⋯⋯⋯⋯⋯⋯⋯⋯⋯⋯⋯⋯⋯⋯⋯⋯ 40
    2.3.2 听力习题设计 ⋯⋯⋯⋯⋯⋯⋯⋯⋯⋯⋯⋯⋯⋯⋯⋯⋯⋯⋯⋯ 42
    2.3.3 听力音频制作 ⋯⋯⋯⋯⋯⋯⋯⋯⋯⋯⋯⋯⋯⋯⋯⋯⋯⋯⋯⋯ 45
    2.3.4 听力词表制作 ⋯⋯⋯⋯⋯⋯⋯⋯⋯⋯⋯⋯⋯⋯⋯⋯⋯⋯⋯⋯ 46

2.4 大语言模型在口语教学中的应用 ⋯⋯⋯⋯⋯⋯⋯⋯⋯⋯⋯⋯⋯⋯ 50

    2.4.1 小组辩论的立论与驳论内容促成 ⋯⋯⋯⋯⋯⋯⋯⋯⋯⋯⋯ 50
    2.4.2 音节接龙语言促成活动设计 ⋯⋯⋯⋯⋯⋯⋯⋯⋯⋯⋯⋯⋯ 52
    2.4.3 听后续说任务设计 ⋯⋯⋯⋯⋯⋯⋯⋯⋯⋯⋯⋯⋯⋯⋯⋯⋯ 53
    2.4.4 "视后续说"任务设计 ⋯⋯⋯⋯⋯⋯⋯⋯⋯⋯⋯⋯⋯⋯⋯ 56
    2.4.5 英语演讲讲稿修改和即兴问答设计 ⋯⋯⋯⋯⋯⋯⋯⋯⋯ 59

2.5 大语言模型在阅读教学中的应用 ⋯⋯⋯⋯⋯⋯⋯⋯⋯⋯⋯⋯⋯⋯ 61

    2.5.1 阅读材料编制 ⋯⋯⋯⋯⋯⋯⋯⋯⋯⋯⋯⋯⋯⋯⋯⋯⋯⋯⋯⋯ 61
    2.5.2 阅读习题设计 ⋯⋯⋯⋯⋯⋯⋯⋯⋯⋯⋯⋯⋯⋯⋯⋯⋯⋯⋯⋯ 64
    2.5.3 交互阅读游戏设计 ⋯⋯⋯⋯⋯⋯⋯⋯⋯⋯⋯⋯⋯⋯⋯⋯⋯ 67
    2.5.4 识别争议焦点的观点促成活动设计 ⋯⋯⋯⋯⋯⋯⋯⋯⋯ 70
    2.5.5 阅读教学思维导图绘制 ⋯⋯⋯⋯⋯⋯⋯⋯⋯⋯⋯⋯⋯⋯⋯ 72
    2.5.6 阅读语篇主题词云图绘制 ⋯⋯⋯⋯⋯⋯⋯⋯⋯⋯⋯⋯⋯ 74

2.6 大语言模型在写作教学中的应用 ⋯⋯⋯⋯⋯⋯⋯⋯⋯⋯⋯⋯⋯⋯ 76

    2.6.1 不同语域同题文本生成 ⋯⋯⋯⋯⋯⋯⋯⋯⋯⋯⋯⋯⋯⋯⋯ 76
    2.6.2 特定难度主题相关词句推荐 ⋯⋯⋯⋯⋯⋯⋯⋯⋯⋯⋯⋯⋯ 78
    2.6.3 典型写作交际场景设计 ⋯⋯⋯⋯⋯⋯⋯⋯⋯⋯⋯⋯⋯⋯⋯ 79
    2.6.4 同义改写语言促成任务设计 ⋯⋯⋯⋯⋯⋯⋯⋯⋯⋯⋯⋯⋯ 82

	2.6.5	典型学生写作样本	83
	2.6.6	议论文和说明文的读后续写设计	86
	2.6.7	英语写作修辞的以续促学设计	90
	2.6.8	视读后续写任务设计	93
	2.6.9	作文自动评阅	96

2.7 **大语言模型在翻译教学中的应用** ··············· 98
    2.7.1 使用目标词翻译的语言促成设计 ··············· 98
    2.7.2 汉英续译任务设计 ··············· 100
    2.7.3 不同翻译策略译例生成 ··············· 103
    2.7.4 口译重要概念解释 ··············· 104
    2.7.5 生成双语术语表抽取 ··············· 108
    2.7.6 口译交传练习设计 ··············· 111

2.8 **大语言模型在词典编纂中的应用** ··············· 113
    2.8.1 学习词典词条生成 ··············· 113
    2.8.2 生成词条配图绘制 ··············· 115

第三章　大语言模型在外语研究中的应用

3.1 文献阅读与评述 ··············· 120
    3.1.1 文献资料查询 ··············· 120
    3.1.2 文献观点提炼 ··············· 121
    3.1.3 同类文献汇总 ··············· 122
    3.1.4 研究趋势挖掘 ··············· 124

3.2 量化数据采集与分析 ··············· 126
    3.2.1 语料收集与标注 ··············· 127
    3.2.2 问卷设计与分析 ··············· 137
    3.2.3 句法语义分析 ··············· 142

  3.2.4 数据的可视化 ·· 150
  3.2.5 量化统计分析 ·· 155
 3.3 质性数据采集与分析 ··· 165
  3.3.1 访谈材料处理 ·· 165
  3.3.2 话语分析标注 ·· 170
  3.3.3 教学研究材料标注 ·· 189
 3.4 语言表达与润色 ··· 200
  3.4.1 近义表达替换 ·· 200
  3.4.2 以语义查词汇 ·· 202
  3.4.3 综合语言润色 ·· 205
  3.4.4 论文格式调整 ·· 209
  3.4.5 投稿修改完善 ·· 214

第四章　结语

 4.1 大语言模型外语教学与研究应用的挑战 ······························ 224
 4.2 大语言模型外语教学与研究应用的机遇 ······························ 226

参考文献 ·· 229

后记 ·· 233

第一章

大语言模型概述

1.1　大语言模型的定义和特点

　　大语言模型（large language model，简称 LLM）是利用深度学习技术训练语言相关数据得到的一种计算系统。当前主流的大语言模型主要指基于"提示语"（prompt）的输入信息，生成词语、代码或其他数据序列的算法系统（Floridi & Chiriatti 2020：684），可用于生成文章、程序代码、图片、音乐、视频等。由于此类模型目前主要使用海量文本数据，即语言数据，进行训练，故而得名。此外，这类模型采用的深度神经网络算法也被用于图片、音视频等多模态数据的训练和应用开发，因而文献中也以"大模型"概称此类人工智能技术。

　　2022 年之后大语言模型领域较具代表性的应用包括：OpenAI 公司的 GPT 大语言模型，包括用于 ChatGPT 聊天机器人的 GPT-3.5、GPT-4、DALL·E 模型；谷歌公司用于 Gemini（之前称 Bard）平台的 LaMDA、PaLM 2、Gemini 模型；Anthropic 公司的 Claude 2 模型等。我国则有百度公司的文心一言大语言模型、科大讯飞公司的星火认知大语言模型、智谱华章公司的智谱清言大语言模型，以及阿里云公司的通义大语言模型。

　　在应用端使用大模型时，普通用户可通过自然语言在软件界面提出需求，大模型则会给出回答或完成任务。大模型能够有效理解人类指令，并根据要求生成相应内容。因此，此轮技术创新又被称为"生成式人工智能"（generative artificial intelligence）。

　　大模型的能力可大致分为语言理解和内容生成两方面。在语言理解方面，大模型较此前的自然语言处理技术进步明显，即便人们用不同的语言风格进行提问，大模型都能抓取要点，正确理解。大模型不仅能够准确地理解简单的问题，对于长篇文档乃至整本书刊，也能有效捕捉大意、提炼要点、交互问答。在文本生成方面，大模型可根据用户指令，生成语句、段落、篇章、程序代码、图片和音视频。用户若连续提问，大模型也可接续回答，从而反馈更符合用户需求的结果。大模型可以胜任各类文案撰写任务，如会议通知、发言提纲、活动新闻、商品广告、营销方案、工作总结、电子邮件等，不一而足。

在语言理解方面，大模型能接受灵活的表达方式，对于不完全精准的提示语有一定的容错能力，有时甚至可以从相对委婉、模糊的指令中准确获取用户需求。在内容生成方面，大模型生成的语句流畅自然，并且可根据用户要求生成不同语言风格的文本，这说明大模型对语言使用的把握游刃有余。此外，在针对给出的文本进行语言学分析时，大模型的表现也可圈可点。

此外，由于大模型所用的海量文本汇集了各个领域的人类知识，因此它可以基于这些知识作出推理，乃至预测。以上所谈的大模型特点使其具备了一定的通用能力，可广泛用于社会生活的方方面面。相关应用的开发方兴未艾，已显示出深入和蔓延之势。从大的应用领域来看，在教育、商业、医疗、法律等方面，大模型几乎无所不包，在学术研究中运用大模型也变得越来越普遍。

研究对象涉及文本或研究方法依赖文本数据的人文社科领域都有大模型的用武之地。语言学和应用语言学、新闻传播学、历史学、法学都较大程度围绕文本开展研究；心理学、社会学、教育学又较常采用访谈、调查等研究方法，其中获得的研究文本也可以利用大模型进行有效挖掘和提炼。

随之而来的是，有关大模型"是敌是友"的讨论不绝于耳（Lin 2023）。在技术的迭代升级和人们的体验深化之后，已经无人能够否认和拒绝大模型的效率和泛在性。当然，大模型不会成为压垮教育和学术的最后一根稻草。人类历史上，每次重大技术变革在引发危机感的同时，往往激励人类更全面、更深刻地认识世界和认识自我。人们更应讨论的是如何善用、巧用大模型，并规避其中的风险。以下，我们将概述大模型在外语教学与研究中的应用思路。

1.2 大语言模型在外语教学与研究中的应用概述

本节所谈的应用概述并非对先行研究和实践的综述，更多是对大模型在外语教学与研究中的应用的设想，因而具有很强的探索性。将大模型应用于外语教学和外语研究实践，从而提高教学和科研效率，为外国语言学及应用语言学学科的提质升级创造了条件，提供了手段。

首先，教学研究中一些原本耗时费力的工作，利用大模型可即刻完成。例如，在听、说、读、写、译相关语言能力的教学准备中，教师可以根据学生的学段和水平、教学话题和课型，利用大模型生成例句（朱奕瑾、饶高琦2023）、语篇、测试练习或其他活动任务。这些教学材料由大模型自动生成，并非直接取自现有出版物，学生也无法通过网络检索直接获得，因而给教学带来了很多原创内容和新鲜元素。比如，我们可以让大模型针对同一主题生成不同语域的多版本学习材料，如正式、非正式两个版本的同题篇章，参见2.6.1节"不同语域同题文本生成"。我们还可以生成包含特定教学要点的文本或练习，如2.6.7节"英语写作修辞的以续促学设计"中，我们要求大模型在生成的文本中包含10处明喻和拟人的修辞用法。在以往的备课环节，这类教学材料相对来说不易获得，但在大模型的助力下，我们可快速生成可用的文本内容，而且通常无须复杂的修改。

教师若对大模型生成内容的主题和质量不满意，可通过调整提示语，明确或细化教学目标，逐步"诱导"出所需内容。比如，我们可在已生成的教学材料基础上，通过修改提示语以增加思辨、思政、跨文化等教学元素。

其次，对于善于学习的学生而言，他们可以将大模型作为贴身的学伴。事实上，已有许多信息技术公司开发出基于大模型的助学应用。学生可以将作文提交给大模型评阅，由模型自动给出反馈，参见2.6.9节"作文自动评阅"。在口译相关任务的训练和准备阶段，学生可参考2.7.4节"口译重要概念解释"、2.7.5节"生成双语术语表抽取"、2.7.6节"口译交传练习设计"，以大大节省译前准备的时间，并有效缓解因时间紧、话题生带来的焦虑，从而显著提升口译练习的质量和效果。随着大模型的普及，学生完全有可能将其作为学习词典来使用，参见2.8节"大语言模型在词典编纂中的应用"的相关案例。

在本书写作过程中，我们较多关注大模型在中国本土外语教学理论（如"产出导向法"和"续理论"）相关教学实践中的应用。同时我们还以大模型的方式对此前基于语料库的数据驱动学习（data-driven learning）教学应用（何安平等 2020）进行了重新设计。

本书展示的教学案例主要集中在外语教学内容的生成方面。大模型灵活的生成能力无疑提升了个性化学习的可操作性。比如，大模型可为不同学习风格、不同水平和不同学习阶段的学生提供风格和难度相适应的语言学习内容。另外，大模型能辅助构建语言交流情境，增强教学活动的沉浸体验。比如，大模型能模拟与母语者进行外语交流的仿真场景。我们甚至可以将外语交流对象设定为苏格拉底、莎士比亚、图灵、联合国秘书长、国际足联主席等具体人物，为对话赋予明确的语境，增强学生使用外语的趣味性和成就感。

在外语研究方面，大模型的出现也必将引发系统性的变革。我们完全有可能构建"大模型语言学方法论"（LLM linguistics methodology）。本书第三章正是在这一思路下的初步尝试。从各节标题——"文献阅读与评述""量化数据采集与分析""质性数据采集与分析""语言表达与润色"——我们不难看出，大模型可以赋能外语研究中的选题挖掘、研究实施、论文写作等多个阶段。对于研究的各个环节，我们都倡导通过"人机协同"（human-AI collaboration）来完成，从而保证研究的方向和价值，切实反映研究者本人的意图。

理论上，大模型可以指出研究空缺，并建议研究选题，然而我们明确反对选题确定环节由机器主导。研究人员仍需通过学习和积累来掌握某个领域或议题的学科背景、核心知识以及现实需求，从学理和实践角度提出真问题，而不能任由人工智能指引方向。我们乐于见到在与大模型的互动交流中，即人机之间的"相互切磋""相互启发"中，获得有价值的选题。

目前来看，大模型在研究数据采集和分析过程中作用明显，是"大模型语言学方法论"的关键应用场景。与外语教学案例显著不同，在本书介绍的外语研究案例中，我们注意到大模型在文本生成之外，其自然语言理解能力体现得较为充分。例如，大模型在解读言外之意时表现出众，它可以较好地识别和标注言语行为、概念隐喻（conceptual metaphor）等内容。另外，大模型在引入相应统计分析模块后，能够出色完成现有语言学实证研究中的多数数据分析任务。

除了 1.1 节介绍的通用大模型外，大模型的衍生科研工具也层出不穷，比如 ChatDOC、Consensus、Connected Papers、Elicit、ResearchRabbit、ScholarAI（参见 3.1.1 节"文献资料查询"）、Scispace、SearchSmart。这些科研工具将大模型与科技文献数据库加以结合，互为补充。大模型有时生成的文献信息不够准确，可借由权威科技文献库进行校正；同时，大模型应用能通过聊天对话的互动方式，进行文献梳理、选题甄别、研究方法比较，还能完成文献管理和文字润色等各项科研任务。在学术文献资料持续增长的背景下，上述工具能帮助我们快速概述文献要义、查询理论概念、确定引文出处等。我们认为，以大模型直接生成学术论文显然违背学术伦理，属于学术不端，但使用大模型完成重复性的数据整理和加工等科研任务属于大模型的合理应用。

1.3 外语教学与研究中的提示工程方法

作为大模型的用户，我们主要通过提示语来实现教学和科研需求。提示语的编写工作被称为"提示工程"（prompt engineering）。提示工程有被神秘化的倾向，这主要是因为提示语的编写具有高度的灵活性。事实上，每个人都有成为最佳提示工程师的潜质。提示语主要服务用户个人和特定任务，因此，编写优质提示语的核心是明确用户需求。只有真正思考清楚个人的任务目标，才有可能编写出明晰的提示语。

互联网上关于提示工程的资料十分丰富，其中常提及"零样本提示"（zero-shot prompting）、"少样本提示"（few-shot prompting）、"思维链提示"（chain-of-thought prompting）等方法。通俗而言，这些提示工程方法可分为"直觉型提示"（intuitive prompting）和"限定型提示"（definitive prompting）两类。我们主张在提示词的编写实践中，可优先选择直觉型提示，将我们想到的问题以日常语径直交由大模型进行回答。如未能得到理想回复，则修改提示语，使之更为具体，可通过列出定义、言明出处、指定格式、设定角色、拆解任务、提供样例等方式帮助大模型理解任务要求。

大模型中已积累了大量的人类知识。经典问题、常识问题、热点问题在训练语料中广泛存在，因而大模型可以较好地理解和回答此类问题。比如，大模型对于"请问翻转课堂的核心理念是什么？""请问乔姆斯基的语言学思想是什么？"等问题，应该能给出教科书般的回复。除了知识性问题之外，大模型的分析、推理能力也相当出色。例如，对于以下问题"请帮忙用符合逻辑的方式对以下四个要点进行排序""" "[1]（1）重塑学科范围；（2）梳理学科脉络；（3）注重本土议题；（4）构建理论框架。"""，并说明排序的理由"，GPT-4 大模型会给出诸如第一项内容为"梳理学科脉络：这是一个自然的起点，因为在任何学科领域进行深入研究之前，首先需要了解其历史、主要理论、关键研究者和过去的研究。这有助于为后续的工作奠定基础……"的回答。对于更细化的教学和研究需求，我们可能需要经过跟大模型多轮对话，才能确定一个定位清晰、输出合理的提示语。在这一过程中，我们可以根据大模型的反馈来调整提示语，借用其中准确合理的用词。我们往往可以通过多个轮次中大模型的欠佳表现进一步了解自身需求，这一试错过程对于完善提示语也很有帮助。完善提示语编写的要领是：（1）厘清需求；（2）模仿借用；（3）追问逼近。

在将某个大模型提示语应用于外语研究时，为保证其标注和分析的有效性，须做好人机标注和分析结果的校验，不可盲目采信大模型生成的内容。对于外语教学和外语研究的从业人员而言，大模型主要扮演"副驾驶"（copilot）的角色。它可以为我们提供飞行建议，但航行目的地应由机长选定，驾驶行为亦应由机长主导。既然将大模型视作人类的智能伙伴，我们在与之交流时，也应与日常人际交流一样，使用清晰、礼貌、得体的措辞。例如，我们常用"Could you create/generate …"等句型来表达我们的指令。因为本书主要聚焦英语教学和英语相关的研究，多数情况下我们采用英文提示语，以便大模型给出有效反馈。

---

[1] OpenAI 官方指南建议可采用三重双引号标示用户内容，将待分析文字和任务说明区分开来。三重双引号这一方式是有效的，可以采用，但它并非强制性的提示表达符号。

第二章

大语言模型在外语教学中的应用

2.1 大语言模型在词汇教学中的应用

2.1.1 学习包含词根 port 的英语单词

◆ 问题与目标

通过词根理解词义并拓展词汇是词汇学习的有效手段。英语词汇 transport（运输）、import（进口）、export（出口）、report（报告）都含有词根 port。port 源自拉丁语，原意为"携带"或"运送"。

我们可以设计练习加深学生对词根 port 构词的理解，并增强对英语构词法的认识。以下是 GPT-4 生成的相关练习。习题题干对某个包含词根 port 的英语单词进行释义，选项中的干扰项为其他含有词根 port 的词汇。

◆ 提示与结果

> 🔊 提示语（有删减）

Please create five multiple-choice questions, each focusing on a different word derived from the root "port". For each question, provide a description of the word for the question and include several answer choices that are correct derivations of "port". Additionally, please provide the correct answer for each question.

适用模型：GPT-4

> ✅ 结果（有删减）

Sure, I'll create five multiple-choice questions focusing on different words derived from the root "port". The root "port" typically relates to carrying or bringing something. Here are the questions, along with their correct answers:

1. Question: This word describes a document that allows individuals to travel between countries. What is it?
   - a) reporter - b) supporter - c) passport - d) transport
   - Correct Answer: c) passport
2. Question: This word refers to the act of carrying goods or people from one place to another, often using vehicles. What is it?
   - a) deport - b) transport - c) report - d) export
   - Correct Answer: b) transport

3. Question: This word is used to describe a person or thing that carries or communicates something. What is it?
   - a) porter - b) portfolio - c) reporter - d) importer
   - Correct Answer: a) porter
4. Question: This term is used for the official process of sending someone out of a country, often for legal reasons. What is it?
   - a) export - b) report - c) deport - d) support
   - Correct Answer: c) deport
5. Question: This word refers to the introduction of goods or services into one country from another. What is it?
   - a) import - b) deport - c) support - d) airport
   - Correct Answer: a) import

These questions cover a range of words derived from the root "port", highlighting their different meanings and applications.

◆ 解读与反思

GPT-4 基于词根 port 设计了五道选择题。我们还可利用相同的方法学习含有词根 port 的其他词汇，例如 airport（机场）、portfolio（作品集）、export（出口）、report（报告）和 support（支持）。通过学习含有相同词根的词汇，学习者可运用构词法知识来记忆新单词，理解英语单词的构造、含义及单词间的内在联系，提高语言学习效率。

2.1.2　围绕特定主题造句成段语言促成设计

◆ 问题与目标

"产出导向法"（production-oriented approach，简称 POA）包括教学理念、教学假设和以教师为中介的教学流程三个部分。其中教学流程由"驱动""促成"和"评价"三个阶段构成，教师在其中扮演关键的中介角色。围绕中心意义组句成段是"产出导向法"语言促成的重要部分，具体方法为教师引导学生选定特定关联词汇进行造句成段，促进基于关联词汇的写作或口语产出。该方法可建立起以功能为中心的目标语言意义关联，带动多个目标词汇的学习和产出，形成更加高效的语言资源功能化促成（邱琳 2017）。

本案例中，我们使用 GPT-4 围绕"克服困难"这一主题自动生成 10 个符合欧洲语言共同参考框架（CEFR，下文简称欧框）B1 水平的词汇，并生成造句成段的任务说明及例子，供教师参考。

◆ 提示与结果

🔊 提示语（有删减）

Please generate 10 interconnected words that are suitable for CEFR B1 level learners and can be used in the context of "overcoming difficulties". These words should encompass themes such as facing difficulties, assessing difficulties, persistence and resilience, and self-encouragement. Following this, design a task that encourages students to write a paragraph about "a memorable experience of overcoming a challenge in their lives", utilizing these 10 words. Additionally, provide an example of writing using these 10 words to demonstrate their application.

适用模型：GPT-4

✅ 结果（有删减）

For CEFR B1 level learners, it's important to choose words that are challenging yet comprehensible. Here are 10 words suitable for the context of "overcoming difficulties":

1. Challenge
2. Persevere
3. Obstacle
4. Resilience
5. Determination
6. Overcome
7. Struggle
8. Courage
9. Triumph
10. Reflect

Task Design:
Title: My Journey Overcoming a Challenge
Instructions:
- Think of a memorable experience where you faced and overcame a challenge.
- Write a paragraph describing this experience.
- Use the following 10 words: challenge, persevere, obstacle, resilience, determination, overcome, struggle, courage, triumph, reflect.
- Focus on how you felt, what the challenge was, and how you overcame it.
- Your paragraph should be between 100–150 words.

Example Paragraph:
"Last year, I faced a significant challenge when I decided to learn a new language. At first, the obstacle of understanding complex grammar seemed insurmountable. However, I was determined to persevere. With determination, I dedicated an hour each day to study. There were times when I wanted to give up, but my resilience kept me going. The struggle with pronunciation was particularly tough, yet I did not lose courage. Slowly, I began to notice improvement and overcame the initial difficulties. When I finally had a conversation with a native speaker, it was a moment of triumph. Looking back, I reflect on this journey with pride, knowing it taught me not just a new language, but also the value of persistence and hard work."

This task and example will help B1 level learners to not only learn new vocabulary but also to practice constructing narratives and expressing personal experiences in English.

◆ 解读与反思

　　GPT-4给出了与"克服困难"相关的10个词汇：challenge、persevere、obstacle、resilience、determination、overcome、struggle、courage、triumph、reflect。教师若需讲授其他主题，则可通过调整提示语获得相应的语言表达，进而引导学生围绕中心意义造句成段，进行语言促成环节。

　　教师应根据学生水平、学习阶段、任务类型对大语言模型提供的语言材料进行相应的编辑。

2.1.3　指定英语词汇的读后续写材料设计

◆ 问题与目标

"续论"语言习得观认为语言是通过"续"习得的，以"续"为中介可以将语言产出与语言理解紧密关联。"续作"是基于续论开发的各种促学语言方法，这些促学方法的共同特征是提供内容不完整的语篇，引导学习者在充分理解语篇的基础上，发挥想象力、模仿原作接续补全语篇内容。续作的输入方式包括视、听、读，输出方式包括续写、续说、续译（王初明 2016，2017）。"读后续写"是以阅读为语言理解方式、以写作为语言产出方式的"续作"，近年来已广泛应用于外语教学与实践。姜琳、涂孟玮（2016）将"读后续写"引入词汇学习领域，发现该任务对二语词汇学习有明显的促学功效。

在实际的读后续写教学过程中，含有指定教学词汇且难度适宜的阅读材料不易获取。但借助大语言模型，我们可以生成符合要求的材料。以下为 GPT-4 基于 10 个教学目标词汇（innovate、diverse、perspective、negotiate、influence、sustainable、contradict、assess、implement、compromise），围绕"绿色能源科学研究"主题生成的读后续写材料。提示语中规定阅读材料的长度约为 300 词，难度符合欧框 B2 水平。

◆ 提示与结果

🔊 提示语（有删减）

Please create a reading-writing integrated continuation task suitable for CEFR B2 level learners, focusing on green energy scientific research. The reading passage should be approximately 300 words long and incorporate the following 10 words: "innovate", "diverse", "perspective", "negotiate", "influence", "sustainable", "contradict", "assess", "implement", and "compromise". After the reading passage, please also provide an example of the continuation writing that extends the topic discussed in the reading passage.

适用模型：GPT-4

> 结果（有删减）

Reading Passage: Innovating for a Greener Tomorrow

In the world of green energy scientific research, the drive to **innovate** is stronger than ever. Researchers from **diverse** backgrounds are coming together to offer a unique **perspective** on sustainable energy solutions. These experts **negotiate** constantly, not just with each other, but with governments, industries, and communities, understanding that their **influence** can shape the future of our planet.

At the heart of their research lies the concept of **sustainable** energy—energy that meets our present needs without compromising the ability of future generations to meet theirs. This sustainable approach often leads to a clash of ideas, as traditional methods **contradict** new scientific findings. For instance, the long-held belief in fossil fuels as the primary energy source is now being challenged by new, cleaner technologies like wind and solar power.

The task to **assess** the viability of these new technologies is complex. Researchers analyze data, consider environmental impacts, and evaluate economic feasibilities. It's a delicate balance, weighing the benefits of new energy sources against the practicalities of **implementing** them on a global scale.

Implementing green energy solutions often requires **compromise**. It involves rethinking how energy is produced, distributed, and consumed. Stakeholders from various sectors must come together, agreeing to make changes that might be initially challenging but are essential for long-term sustainability.

The journey towards a greener future is paved with challenges and opportunities. As we navigate this path, the collective effort to innovate, negotiate, and implement sustainable practices will determine the health of our planet for generations to come.

Writing Task: Extending the Topic

Task: Based on the reading passage, write a 250-word essay discussing the role of government policies in supporting green energy initiatives. Consider how governments can influence the shift towards sustainable energy sources and what compromises might be necessary to balance economic and environmental concerns.

◆ 解读与反思

　　GPT-4 为读后续写任务生成的阅读材料篇幅适中，词汇复杂度符合要求，并成功将 10 个教学目标词汇全部融入阅读材料。但观察生成文本发现，大模型严格按照指定词汇在提示语中的顺序生成文本，行文难免生硬。后续我们可以进一步精调提示语，要求大模型不拘泥于给出词汇的顺序，生成逻辑衔接更为自然的语篇。

2.1.4　fare 和 fee 的词义辨析

◆ 问题与目标

对于外语学习者来说，分辨近义词的意义差异、根据语境准确选择适宜词汇存在一定的困难。在英语中，fare 和 fee 都有"费用"的意思，但二者所指的费用类型有明显区别。fare 通常指的是与交通运输相关的费用（如搭乘公共交通的费用）以及特定活动（如电影等）的费用；而 fee 指的是与服务、许可证、会员资格、注册等有关的费用，通常支付给个人、组织机构、政府部门等。

为帮助学习者辨析近义词的意义差异，我们可以使用大语言模型生成练习，引导学生结合语境对比分析词义。本案例使用 GPT-4 生成辨析 fare 和 fee 词义差异的选词填空题。

◆ 提示与结果

🔊 提示语（有删减）

Please create 20 fill-in-the-blank questions (using "fare" or "fee") to distinguish the meanings of "fare" and "fee", and provide the answers along with explanations.

适用模型：GPT-4

✅ 结果（有删减）

The ___ for the subway ride is $2.50.
Answer: fare
Explanation: When referring to the cost of transportation, "fare" is used.
The entrance ___ to the museum is $15.
Answer: fee
Explanation: Charges for services or admission to facilities/events are called "fees".
How did you ___ on your math exam?

Answer: fare
Explanation: "Fare" can also mean to perform in a specified way in a particular situation.
The doctor's consultation ___ is not covered by my insurance.
Answer: fee
Explanation: Charges for professional services are typically called "fees".
The taxi ___ from the airport to the hotel can be quite expensive.
Answer: fare
Explanation: The cost for transportation services like taxis is termed "fare".
…

◆ 解读与反思

尽管提示语没有提供任何关于 fare 和 fee 意义差异的具体信息，大模型依然准确生成了符合词义的练习，并覆盖了目标词汇的典型使用场景，如与 fare 有关的地铁费用（fare for the subway）、出租车费用（taxi fare）等，以及与 fee 有关的博物馆门票费用（fee to the museum）、咨询费（consultation fee）等。此外，大模型还根据题干语境提供了答案解析，概括出了不同词汇对应的不同费用类型，有助于学生举一反三。

不过，大模型生成的练习还涉及了目标词汇的非"费用"含义，如 fare 表示"进展"（How did you fare on your math exam?）。如果希望大模型只考虑目标词的特定义项，提示词需对此做出明确说明。

2.1.5　cloth、cloths、clothing 和 clothes 的词义辨析

◆ 问题与目标

另一个近义词辨析的例子是 cloth、cloths、clothing 和 clothes 这四个形式相似、语义关联的词。正因为它们在形式和意义上的相似性，学习者常常不能准确分辨它们的差别。笼统来说，cloth 和 cloths 可归为布；clothing 和 clothes 可归为衣。具体来说，cloth 指的是布料，常出现在 cotton cloth（棉布）、silk cloth（丝绸）等搭配中表示布料的质地，或出现在 cloth napkin（餐巾布）、table cloth（桌布）等搭配中表示布料的用途。cloths 是 cloth 的复数形式，cloth 在泛指布料时不可数，而在指不同类型、不同种类或裁开后的布料时可数。clothing 统指衣服和服饰，为不可数名词。clothes 这一复数形式则指具体的衣服。

本案例使用 GPT-4 针对 cloth、cloths、clothing 和 clothes 的词义辨析设计改错题，引导学生关注语境，从而帮助学生理解、掌握相应词语意义差异。

◆ 提示与结果

🔊 提示语（有删减）

Please create 20 error-correction questions to distinguish the meanings of "cloth", "cloths", "clothing" and "clothes". Please provide answers and explanations.

适用模型：GPT-4

✅ 结果（有删减）

Incorrect: The store sells beautiful cloth.
Correct: The store sells beautiful clothes.
Explanation: "Clothes" refers to items we wear.

Incorrect: I spilled wine on the table cloth.
Correct: I spilled wine on the table cloth.
Explanation: No correction needed. "Cloth" refers to the piece of fabric covering the table.

Incorrect: Can you pass me the clothes to wipe the dishes?
Correct: Can you pass me the cloths to wipe the dishes?
Explanation: We use multiple pieces of fabric (cloths) to wipe dishes, not wearable items.

Incorrect: He's donating his old clothing to charity.
Correct: He's donating his old clothes to charity.
Explanation: "Clothes" refers to individual items of clothing.

Incorrect: I need to buy a piece of clothes for my sewing project.
Correct: I need to buy a piece of cloth for my sewing project.
Explanation: We're referring to a single piece of fabric here.

Incorrect: The artist uses different cloths for her projects.
Correct: The artist uses different cloths for her projects.
Explanation: No correction needed. "Cloths" indicates multiple pieces of fabric.
……

◆ 解读与反思

  大模型生成的改错题囊括了四个目标词的典型使用场景，能够帮助学生学习并巩固词语的意义差异。但值得注意的是，生成结果中也存在一些问题。例如第一题中，题干有限的语境并不足以判断店内销售的是服装还是布料，因而答案的依据并不充分；展示出的最后一题中，答案解析认为这里的 cloths 表示"多块布料"，但从语境来看，"多种材质的布料"这一意义同 different 一词的修饰才更为契合。我们可以考虑修改提示语，让大模型生成的题干包括更丰富、精确的语境。本案例的结果也再次说明教师需仔细检查大模型生成的内容，必要时应对材料进行修改。

2.1.6 suggest 和 advise 的用法辨析

◆ 问题与目标

  前文中的两个案例主要涉及语义相似的近义词辨析，但近义词在语用上的差异同样会给学习者造成困难。学习者在不了解词汇语用差异的情况下，可能会产出不合时宜甚至不礼貌的表达，从而影响交际效果。

  比如，尽管都用于给他人提出意见，英语词汇 suggest 和 advise 一般应用于不同的交际场景。suggest 多应用于非正式场合，对于建议者、建议的权威性无具体的限定；advise 则多应用于正式或官方场合，强调建议者具有权威的专业知识，并明确告知建议接收者应该怎样去做。二者的句法模式也有所区别。suggest 主要用在 suggest sth.、suggest doing、suggest that sb. (should) do sth. 等结构中，advise 则主要用在 advise sth.、advise sb. to do sth.、advise that sb. should do sth. 等结构中。既往教学实践表明，学习者使用 suggest 和 advise 时，在语用场景和句法结构上都容易出现错误（何安平等 2020）。

为帮助学习者进一步巩固 suggest 和 advise 语用场景和句法结构上的差异，我们使用 GPT-4 围绕 suggest 和 advise 的用法区别生成了判断练习和答案解析。

◆ 提示与结果

◁)) 提示语（有删减）

Please provide five exercises to distinguish the difference between "suggest" and "advise" when expressing recommendations, with more than half of the sentences containing incorrect usage between "suggest" and "advise" concerning the wrong situations (e.g. the doctors advised the patients (instead of using "suggest")) or grammatical mistakes (e.g. suggest sb. to do sth.). Then have students judge the correctness of these 20 example sentences. Please provide comments on the correctness of these five example sentences.

适用模型：GPT-4

✓ 结果（有删减）

Exercise: Correct or Incorrect?
Instructions: Determine whether the usage of "suggest" or "advise" in each sentence is correct or incorrect.

1. Sentence: "I suggest you to take a day off." (Incorrect)
- Comment: The correct form is "I suggest taking a day off" or "I suggest that you take a day off."
2. Sentence: "My lawyer advised me to plead guilty." (Correct)
- Comment: This is a correct usage of "advise" as it's a recommendation from a professional.
3. Sentence: "She suggests that we should go to the new restaurant." (Correct)
- Comment: This is a proper use of "suggest".
4. Sentence: "The doctor advised me that I should eat healthier." (Correct)
- Comment: "Advised" is correctly used here as it's a professional recommendation.
5. Sentence: "I advise taking a break if you're tired." (Correct)
- Comment: "Advise" is correctly used as a suggestion here.

◆ 解读与反思

　　大语言模型围绕 advise 和 suggest 的使用差异生成了习题，题干中的场景能够充分体现二者的不同语用场景，如 advise 常与权威性和专业建议（如医生、律师等）共现，suggest 则常用于相对日常和非正式的场景（如朋友的建议）。同时，大模型在生成习题时也关注到了如 suggest sb. to do sth. 等语法错误的问题。教师可简单修改提示语，将这类练习应用到其他一义多形现象，如区分 affect 和 effect、imply 和 infer、lie 和 lay 等。

2.1.7　get around 的熟词生义问题

◆ 问题与目标

　　学习者掌握多义词或短语中不同义项的程度往往存在差异。即便词或短语本身十分常见，学习者也可能只是熟悉其中的一部分用法，而对低频或引申的意义相对陌生。比如短语 get around 既有"四处走动"（move from place to place）的基本意思，也有如克服（问题/困难）（get around the problem）、避开（某事或某人）（get around the rules）等引申义。为帮助学习者准确理解熟词生义，我们可以使用大语言模型生成语义辨析练习，锻炼学习者根据语境识别目标短语不同意义的能力。

◆ 提示与结果

> 🔊 提示语（有删减）

Please design multiple-choice questions to distinguish all the meanings of the phrase "get around". Please provide the answers along with explanations.

适用模型：GPT-4

> ✅ 结果（有删减）

Here are multiple-choice questions to distinguish the different meanings of the phrase "get around":
1. Which of the following best describes the meaning of "get around" in the sentence "She finds it difficult to get around since she broke her leg"?

A) Avoid doing something B) Become well-known C) Move or travel from place to place D) Spread a rumor or information

Answer: C) Move or travel from place to place

Explanation: In this context, "get around" means to move or travel from one place to another, particularly in relation to mobility.

2. Which meaning does "get around" have in the sentence "It's hard to get around the strict regulations imposed by the government"?

   A) Avoid or bypass something B) Share news or information C) Visit multiple places D) Become famous or popular

   Answer: A) Avoid or bypass something

   Explanation: Here, "get around" means to find a way to avoid or bypass certain obstacles or challenges, in this case, the strict regulations.

3. In the sentence "News of the celebrity's secret marriage quickly got around", what does "get around" mean?

   A) To travel to various places B) To circumvent an obstacle C) To spread or circulate D) To postpone or delay

   Answer: C) To spread or circulate

   Explanation: In this context, "get around" means that the news spread or circulated among people.

4. Which of the following best describes "get around" in the sentence "Despite his initial reluctance, we managed to get around him and he agreed to our proposal"?

   A) Move from one location to another B) Persuade or convince C) Share a secret D) Go on a tour

   Answer: B) Persuade or convince

   Explanation: Here, "get around" means to persuade or convince someone to change their mind or opinion.

5. In the context of "It's hard to get around the fact that he lied to us", what does "get around" signify?

   A) Visit different places B) Ignore or deny something obvious C) Share a piece of information D) Persuade someone

   Answer: B) Ignore or deny something obvious

   Explanation: In this scenario, "get around" means to ignore or deny an evident fact or reality.

◆ 解读与反思

从题目设计的角度看，大语言模型所生成的单项选择题涵盖了 get around 的几个主要生义，包括题目3中"传播、散播"的意思（News got around）以及题目4中"劝说"的意思（get around him）；此外，大语言模型为题目提供了充足的语境以便学生进行意义的判断。从题目难度来看，大语言模型给出的 get around 例句难度适中；在具体操作过程中，教师可修改提示语，引导大语言模型根据欧框标准来生成难度符合教学需求的题目。

2.1.8 拼写练习的设计

◆ 问题与目标

对于英语学习者而言，拼写错误是一个常见的问题。拼写错误往往由同音异形词的混淆、不符合常规的拼读规则以及不发音字母的存在等因素引起。英语学习者由于受到第二语言能力的限制和母语负迁移的影响，在拼写上容易出错，因此针对性的拼写练习显得尤为重要。下面是 GPT-4 设计的 10 个单词的拼写练习（accommodate、acknowledgment、liaison、maintenance、necessary、privilege、pronunciation、recommend、separate、threshold），每个练习题目包括三个常见的错误拼写方式和一个正确答案，并且展示了单词的使用语境。

◆ 提示与结果

> 🔊 提示语（有删减）

Design a context-based multiple-choice spelling quiz for the words "Accommodate", "Acknowledgment", "Liaison", "Maintenance", "Necessary", "Privilege", "Pronunciation", "Recommend", "Separate" and "Threshold". For each word, include a sentence that provides context (CEFR B2 level), followed by four answer choices: the correct spelling and three commonly misspelled versions by EFL learners.

适用模型：GPT-4

✓ 结果（有删减）

1. Accommodate
   - Sentence: "The hotel was able to _____ all of our requests for additional pillows."
   - Choices:
   a) Accomodate b) Accommodate c) Acommodate d) Acomodate

2. Acknowledgment
   - Sentence: "She received an _____ from the company for her innovative ideas."
   - Choices:
   a) Acknowledgment b) Acknowledgement c) Acnowledgment d) Acnowledgement

3. Liaison
   - Sentence: "He acted as a _____ between the two departments to ensure smooth communication."
   - Choices:
   a) Liaison b) Liason c) Liasion d) Leaison

4. Maintenance
   - Sentence: "Regular _____ is essential to keep the machine running efficiently."
   - Choices:
   a) Maintainance b) Maintenence c) Maintenance d) Maintanence

5. Necessary
   - Sentence: "It is _____ to have a passport when traveling internationally."
   - Choices:
   a) Necesary b) Neccessary c) Necessary d) Neccesary

6. Privilege
   - Sentence: "Having access to advanced education is a _____ that not everyone has."
   - Choices:
   a) Priviledge b) Privilege c) Privelege d) Privelige

7. Pronunciation
   - Sentence: "Her _____ of the foreign words impressed everyone."
   - Choices:
   a) Pronounciation b) Pronunciation c) Prononciation d) Pronuncation

8. Recommend
   - Sentence: "I highly _____ this book to anyone interested in history."
   - Choices:
   a) Recomend b) Recommend c) Reccommend d) Reccomend
9. Separate
   - Sentence: "Please _____ the recyclable materials from the non-recyclables."
   - Choices:
   a) Seperate b) Separate c) Separrate d) Separete
10. Threshold
    - Sentence: "He stepped over the _____ and entered the room."
    - Choices:
    a) Treshold b) Threshold c) Threshhold d) Threshould

◆ 解读与反思

GPT-4 针对一些英语学习者常见的拼写错误，根据提示语中的 10 个单词设计了拼写练习，这些错误涉及重复字母的使用（如 necessary 的错误拼写 necesary 或 neccessary），字母顺序的错误（如 separate 的错误拼写 seperate），额外字母的添加（如 acknowledgment 的错误拼写 acknowledgement）等方面，可有效帮助学习者识别和纠正常见的拼写误区。教师可进一步修改提示语，以引导大语言模型生成更加符合教学需求的拼写练习。

2.2 大语言模型在语法教学中的应用

2.2.1 读后续写关系从句学习材料设计

◆ 问题与目标

"读后续写"以"续"为中介，提供内容不完整的语篇输入，引导学习者模仿原作接续补全语篇内容，将语言产出与语言理解紧密关联，产生拉平效应（王初明 2016，2017）。既往研究表明，基于"续论"语言习得观开展"读后续写"任务，可有效促学复杂的英语句子结构，如王启、王初明（2019）的研究验证了以续促学英语关系从句的有效性。在进行读后续写的教学活动时，可以利用大语言模型,将英语关系从句融入阅读材料,进而设计出读后续写的任务。以下为 GPT-4 围绕"环境保护"主题所设计的读后续写任务,提示语要求 GPT-4 在 200 词左右的阅读材料中嵌入 10 个英语关系从句,且确保难度符合欧框 B2 水平,并需提供续写示例。

◆ 提示与结果

🔊 提示语(有删减)

Please create a reading-writing integrated continuation task suitable for CEFR B2 level learners, focusing on environmental protection. The reading passage should be approximately 200 words long and incorporate 10 relative clauses (please highlight these clauses). After the reading passage, please also provide an example of the continuation writing that extends the topic discussed in the reading passage.

适用模型:GPT-4

✅ 结果(有删减)

Embracing Green Solutions: A Community's Journey

In a small town, **which is nestled in the heart of the mountains**, the residents have embarked on a remarkable journey towards sustainability. This community, **whose dedication to environmental protection is inspiring**, has adopted several green initiatives. One of the most successful projects is the community garden, **where people, who are passionate about organic farming, come together**. The garden not only provides fresh produce but also serves as a hub for learning and sharing techniques **that help in reducing the carbon footprint**. The local school, **which has integrated environmental education into its curriculum**, plays a pivotal role in this transformation. Students, **who learn about renewable energy sources**, are encouraged to think critically about environmental issues. One student, **who designed a rainwater harvesting system**, exemplified the kind of innovation **that the school fosters**.

Another notable effort is the town's recycling program, **which has significantly reduced waste**. The program, **which is run by volunteers**, ensures that materials, **which would otherwise end up in landfills**, are properly recycled. A local entrepreneur, **who started a business that turns recycled plastic into art**, demonstrates how recycling can have economic benefits too.

This town's journey, **which is a testament to the power of community action**, shows that small steps can lead to significant changes. It reminds us that everyone, **who is committed to the cause**, can make a difference in protecting our environment.

Continuation Writing Task
Prompt: Imagine you are a member of this community and have been inspired by the initiatives mentioned in the reading passage. Write a proposal for a new environmental project or improvement to an existing one. Your proposal should be approximately 200 words and include:
1. A clear description of the project or improvement.
2. How it will benefit the community and the environment.
3. Steps for implementation.
4. Possible challenges and solutions.
Use the initiatives in the passage as a reference for your ideas, and remember to include relative clauses to enhance your writing.

◆ 解读与反思

GPT-4很好地基于"环境保护"主题设计了读后续写任务,并且在读后续写的阅读材料中嵌入了至少10个英语关系从句。GPT-4所挑选的英语关系从句主要包含由which、who和that引导的限定性关系从句和由which引导的非限定性关系从句。从阅读材料的长度和难度来看,也符合提示语中200词、欧框B2水平的要求。在读后续写的教学实践中,设计合适的读后续写材料是一大难题,借助大语言模型,教师可以定制符合教学目标的读后续写材料,提高效率,并可进行"人工智能后编辑",进一步调整大语言模型的生成内容以符合教学要求。

2.2.2　读后续写虚拟语气学习材料改编

◆ 问题与目标

基于"续论"语言习得观开展"读后续写"任务,可有效促学二语语法结构。辛声(2017)将强化读后续写(如在阅读材料中明确提示相应的语法结构)作为促学语言的续作,应用于虚拟语气的教学中,发现强化读后续写可有效增强协同效应，促进英语学习者对虚拟语气的习得。以下为 GPT-4 针对虚拟语气设计的强化读后续写任务。对于输入的英语故事（取自 Flora Annie Steel 的 *The Bogey-Beast*），GPT-4 智能地生成阅读材料，将虚拟语气的使用融入其中，并以此设计读后续写任务。

◆ 提示与结果

🔊 提示语（有删减）

Please create a reading-writing integrated continuation task suitable for CEFR B2 level learners, based on the story of *The Bogey-Beast*. The reading segment should be around 300 words, embedding 10 examples of the subjunctive mood, which should be clearly marked. Additionally, include a continuation writing sample that logically extends the story's theme from the reading passage.

Here is the story of *The Bogey-Beast*:
There was once a woman who was very, very cheerful, though she had little to make her so; for she was old, and poor, and lonely. She lived in a little bit of a cottage and earned a scant living by running errands for her neighbours, getting a bite here, a sup there, as reward for her services. So she made shift to get on, and always looked as spry and cheery as if she had not a want in the world.
Now one summer evening, as she was trotting, full of smiles as ever, along the high road to her hovel, what should she see but a big black pot lying in the ditch!
"Goodness me!" she cried, "that would be just the very thing for me if I only had something to put in it! But I haven't! Now who could have left it in the ditch?"
And she looked about her expecting the owner would not be far off; but she could see nobody.
"Maybe there is a hole in it," she went on, "and that's why it has been cast away. But it would do fine to put a flower in for my window; so I'll just take it home with me."
And with that she lifted the lid and looked inside. "Mercy me!" she cried, fair amazed. "If it isn't full of gold pieces. Here's luck!"
And so it was, brimful of great gold coins. Well, at first she simply stood stock-still, wondering if she was standing on her head or her heels. Then she began saying:
"Lawks! But I do feel rich. I feel awful rich!"

…

适用模型：GPT-4

结果（有删减）

Reading Segment with Subjunctive Mood Examples Marked

There was once a woman, cheerful and spry, living alone in her quaint cottage. Despite her meager means, her heart was always full of hope. As twilight caressed the sky one evening, she stumbled upon a curious sight—a large, mysterious pot lying abandoned in a ditch.

"Goodness gracious!" she exclaimed. "If **only I had** something to cook in it! But alas, my cupboard is bare."

She pondered over who might have left such a pot there, her eyes scanning the empty road.

She mused, "Perhaps there's a hole in it, and that's why it's been discarded. But, it would serve well to hold a flower for my window. **I wish it were** mine." With that, she approached and lifted the lid, peering inside.

To her astonishment, it was brimming with gold coins!

"Mercy on me!" she gasped, her eyes widening. "**I had never imagined** finding such fortune. **If it were** possible, I'd be the richest woman in town!" The coins shone brightly, casting a golden glow on her astonished face.

She thought, "Had I **known** about this earlier, I **would have come** here straightaway. It's **essential that I take** this treasure home." Struggling to lift the heavy pot, she tied it to her shawl, dragging it behind her like a wagon.

As she trudged along, she mused, "If **only my neighbors were** here to see this. They'd be as surprised as I am!" Lost in thought, she didn't notice the pot's contents magically transforming with each stop she made, first into silver, then iron, and finally into a mere stone.

Upon reaching her cottage, she untied the shawl, only to find a stone. "Had I **not seen** it myself, I'd **never believe** it," she sighed, yet her spirits remained undaunted. "It's **crucial that I find** a use for this stone now."

Just then, the stone morphed into a creature—large and whimsical. It pranced away, leaving her in awe. "How **fortunate I am to have witnessed** such magic," she beamed. "I **should tell** everyone about this Bogey-Beast!"

Writing Task: Continuation

Continuing from the story's theme, imagine what happens the next day after the woman's encounter with the Bogey-Beast. Write a continuation of the story, focusing on how the woman's experience with the magical creature affects her life and her interactions with her neighbors. Consider introducing new characters or magical elements that tie back to the theme of unexpected fortune and transformation. Your writing should be approximately 300–400 words. Remember to incorporate descriptive language and dialogue to bring your story to life.

◆ 解读与反思

GPT-4 较好地基于故事 *The Bogey-Beast*，针对虚拟语气设计了读后续写任务。在设计的阅读材料中，GPT-4 智能地去掉了结尾部分，并将原文改写为欧框 B2 水平的 300 词阅读材料；GPT-4 选取的虚拟语气包括现实的虚拟（If only I had、If it were）、愿望的虚拟（I wish it were）、过去的虚拟（Had I known、I had never imagined）和命令的虚拟（I should tell）；此外，该读后续写任务引导学生想象女主角在与 Bogey-Beast 相遇后第二天发生的事情来续写故事，并鼓励学生引入新角色。对于大语言模型生成的内容，教师可采取"人工智能后编辑"的方式进一步修改，使其生成内容符合教学任务要求。

2.2.3 宾语补语的用法

◆ 问题与目标

宾语补语（object complement）是英语语法中的重要概念，它通常用于补充说明宾语的性质、状态或身份等，提供关于宾语的额外信息。宾语补语可以是名词、形容词、动词不同形式、介词短语等。以下为 GPT-4 从形容词作宾补、现在分词作宾补和不定式作宾补三个角度出发设计的宾语补语改写练习题。

◆ 提示与结果

🔊 提示语（有删减）

Please create a series of 15 exercises designed for CEFR B2 level English learners, with a focus on the use of object complements. Distribute these exercises across three categories of object complements: adjectives, present participles and infinitive phrases, with five exercises per category. Each exercise should present a base sentence along with specific guidance for modifying it to include an object complement. This guidance could be in forms like "revise using an adjective as the object complement" or "use a past participle to describe the object's state". Additionally, include separate answers for each exercise.

适用模型：GPT-4

结果（有删减）

Exercises on adjectives as object complements
1. Base Sentence: "They considered the proposal."
   - Guidance: Revise using an adjective as the object complement to describe the proposal.
   - Answer: "They considered the proposal beneficial."
2. Base Sentence: "We found the movie."
   - Guidance: Use an adjective to describe the movie.
   - Answer: "We found the movie fascinating."
   …

Exercises on present participles as object complements
1. Base Sentence: "I saw her."
   - Guidance: Add a present participle to describe what she was doing.
   - Answer: "I saw her running."
2. Base Sentence: "They heard the dog."
   - Guidance: Use a present participle to describe the dog's action.
   - Answer: "They heard the dog barking."
   …

Exercises on infinitive phrases as object complements
1. Base Sentence: "I want you."
   - Guidance: Add an infinitive phrase to describe your intention.
   - Answer: "I want you to stay."
2. Base Sentence: "They expected us."
   - Guidance: Use an infinitive phrase to describe what they expected from us.
   - Answer: "They expected us to participate."
   …

◆ 解读与反思

大语言模型设计出了形容词作宾补、现在分词作宾补和不定式作宾补三类宾语补语改写练习题，并给出了参考答案。提示语在编写时有意引导大语言模型生成创造性的改写练习，如形容词作宾补练习的第一题"他们认为这个提案/提议……"（They considered the proposal ...），要求学生使用形容词作宾补去描述"提案/提议"（proposal）。这种创造性的改写练习不仅可有效帮助学生学习宾语补语，还可通过补全的方式帮助学生练习宾语的搭配。

2.2.4 过去完成体和过去完成进行体的区分

◆ 问题与目标

在英语中，过去完成体和过去完成进行体都用于描述发生在过去某一特定时刻的动作或状态。过去完成体（"had+过去分词"）用于描述在过去某一时间点或过去动作之前已经完成的动作，强调动作在过去的完成；过去完成进行体（"had been+现在分词"）用于描述在过去某一时间点或过去动作之前正在进行的动作或状态，强调动作在过去的持续。由于中文没有明确的时态结构，因此对于中国英语学习者来说，过去完成体和过去完成进行体难区分，易混淆，常误用。为了有效引导学生学习两个时态，可通过大语言模型生成相关练习促进习得。以下为GPT-4生成的10道关于过去完成体和过去完成进行体的语法填空题。

◆ 提示与结果

🔊 提示语（有删减）

Please create 10 grammar gap-fill questions that guide students to modify the base form of verbs in given sentences using "past perfect" or "past perfect continuous" tense. Please provide the answers and explanations. Please ensure that the vocabulary used in the practice questions is at CEFR B2 level. Please also ensure the gap-fill questions are about campus life.

适用模型：GPT-4

✅ 结果（有删减）

1. By the time I reached the lecture hall, the lecture _____ (already / start).
   Answer: had already started
   Explanation: The action of the lecture starting took place before another past action (reaching the lecture hall), so we use the past perfect tense.
2. She couldn't find her notebook because she _____ (leave) it in the library the day before.
   Answer: had left
   Explanation: The action of leaving the notebook in the library happened before another past event (her trying to find it), so we use the past perfect tense.
3. They _____ (wait) in the cafeteria for over an hour when we finally met them.
   Answer: had been waiting
   Explanation: This emphasizes a continuous action that was happening in the past over a period of time before another action occurred.
4. The students were upset because they _____ (not / receive) their exam results by the end of the semester.
   Answer: had not received
   Explanation: The action of not receiving the exam results occurred before the students felt upset, so we use the past perfect.
5. When I ran into Jake outside the library, he looked tired because he _____ (study) all night.
   Answer: had been studying
   Explanation: This suggests a continuous action in the past that had an impact on a later event or moment.
6. The football team celebrated because they _____ (win) the inter-college championship.
   Answer: had won
   Explanation: The action of winning took place before the celebration, so we use the past perfect tense.
7. I was surprised to see that the auditorium was empty; I thought the play _____ (already / begin).
   Answer: had already begun
   Explanation: The thought (or expectation) of the play having started happened before the realization that the auditorium was empty.

8. Sarah was well-prepared for her presentation because she _____ (practice) it many times.
   Answer: had practiced
   Explanation: The action of practicing multiple times happened before the actual presentation.
9. By the time the students arrived for the workshop, the instructor _____ (already / set up) all the materials.
   Answer: had already set up
   Explanation: The action of setting up took place before the students' arrival.
10. When the students started their project, they realized they _____ (not / understand) the guidelines properly.
    Answer: had not understood
    Explanation: The realization of not understanding the guidelines came after they started the project, indicating that the misunderstanding occurred in the more distant past.

◆ 解读与反思

大语言模型较为准确地就过去完成体和过去完成进行体设计了 10 道语法填空题，这些题目不仅涉及时态，还涉及 already 的用法（题目 1）、否定句（如题目 4）等，并且每道题目都给出了正确的答案和较为详细的解释。此外，大语言模型还根据提示语将练习题的主题设定为"校园生活"，将用词难度控制在欧框 B2 水平。相较于传统的人工编制与时态有关的语法填空题目的做法，大语言模型可在几秒钟的时间内高效、高质量地生成题目，并可进行定制，控制题目的主题和难度，这显示了大语言模型的优势。对于生成的结果，教师可采取"人工智能后编辑"的方式进一步加工，以符合语法教学的需求。

2.2.5　否定副词引导的部分倒装句学习

◆ 问题与目标

在英语中，倒装句指的是谓语动词或助动词位于主语之前的句子结构，这种特殊的语序可以起到强调、修饰的作用。倒装句可分为完全倒装（整个谓语置于主语之前）和部分倒装（只有 be 动词／助动词／情态动词置于主语之前）。

对于否定副词引导的部分倒装句（partial inversion with negative adverbials），主要的结构有"否定副词+be 动词/助动词/情态动词＋主语＋谓语"的形式，主要的否定副词（短语）包括 never、seldom、hardly、scarcely、not only、in no way、at no time、under no circumstances 等。对于中国英语学习者来说，由于中英文之间的明显差异，学生会认为否定副词引导的部分倒装句与中文逻辑不符，造成困扰。在实际应用过程中，学生会出现完全倒装与部分倒装混淆、be 动词/助动词/情态动词和主语之间的顺序错误、时态错误等问题，因此有必要对否定副词引导的部分倒装句加强练习。以下为 GPT-4 生成的 20 道关于否定副词引导的部分倒装句的改写练习题。

◆ 提示与结果

提示语（有删减）

Please create 20 questions for transforming the original sentences into "partial inversion sentences with negative adverbials". Please provide the negative adverb and the original sentence, and instruct students to rewrite it as a partial inversion sentence. Also, provide the answers. Please ensure that each question uses a different negative adverb and that the vocabulary difficulty level in the exercises is kept at the CEFR B2 level.

适用模型：GPT-4

结果（有删减）

Instructions: Rewrite the given sentences using the provided negative adverb as a partial inversion sentence.

Negative Adverb: Never
Original Sentence: I have seen such a magnificent view.
Answer: Never have I seen such a magnificent view.

Negative Adverb: Rarely
Original Sentence: She gets up early on weekends.
Answer: Rarely does she get up early on weekends.

Negative Adverb: Seldom
Original Sentence: John forgets his keys.
Answer: Seldom does John forget his keys.

Negative Adverb: Hardly ever
Original Sentence: They go to the cinema.
Answer: Hardly ever do they go to the cinema.

Negative Adverb: Not only
Original Sentence: She sings beautifully, but she also dances gracefully.
Answer: Not only does she sing beautifully, but she also dances gracefully.

Negative Adverb: On no account
Original Sentence: You should touch the wires.
Answer: On no account should you touch the wires.

Negative Adverb: In no way
Original Sentence: Is he involved in the scam.
Answer: In no way is he involved in the scam.

Negative Adverb: At no time
Original Sentence: Did they consider quitting the project.
Answer: At no time did they consider quitting the project.

Negative Adverb: Nowhere
Original Sentence: Have I found such delicious sushi.
Answer: Nowhere have I found such delicious sushi.

Negative Adverb: Not until
Original Sentence: She saw the results did she believe in the method.
Answer: Not until she saw the results did she believe in the method.

……

◆ 解读与反思

大语言模型所生成的关于否定副词引导的部分倒装句的改写题大部分质量相当不错；根据提示语的要求，20 道练习题使用了不同的否定副词，包括 never、rarely、seldom 等，从而引导学生积累并习得可应用于部分倒装的不同否定副词或短语。但值得注意的是，大语言模型生成的部分倒装句虽符合语法要求，但并不是常见的语言用法，因此教师应采取"人工智能后编辑"的方式，对生成结果做进一步修改。

2.2.6 主谓一致问题

◆ 问题与目标

主谓一致是英语语法中的基本原则，要求句子中的主语和谓语在人称和数上必须保持一致。主谓一致问题是英语学习者时常出现的问题，尤其是在集体名词作主语、带有介词短语的主语、并列主语、不定代词作主语等情况下，容易出现主谓不一致的问题。以下为 GPT-4 针对主谓一致问题设计的改错练习。

◆ 提示与结果

提示语（有删减）

Please create a set of 20 error correction exercises focusing on subject-verb agreement issues in English. Provide each exercise with its corrected version and detailed explanations for the corrections.

适用模型：GPT-4

结果（有删减）

Exercise 1
Incorrect: She go to the gym every morning.
Corrected: She goes to the gym every morning.
Explanation: The verb should agree with the singular subject "she". The correct form is "goes".

Exercise 2
Incorrect: The dogs barks at night.
Corrected: The dogs bark at night.
Explanation: The plural subject "dogs" requires the base form "bark" instead of the singular form "barks".

Exercise 3
Incorrect: John and Mary is a happy couple.
Corrected: John and Mary are a happy couple.
Explanation: When two subjects are connected by "and", they form a plural subject, so the verb should be "are" instead of "is".

Exercise 4
Incorrect: The list of items are on the table.
Corrected: The list of items is on the table.
Explanation: The subject "list" is singular, even though it refers to multiple items. Therefore, the singular verb "is" should be used.

Exercise 5
Incorrect: Either the cats or the dog need to be fed.
Corrected: Either the cats or the dog needs to be fed.
Explanation: When "either … or …" is used, the verb agrees with the nearer subject. In this case, "dog" is singular, so "needs" is the correct form.

Exercise 6
Incorrect: The number of students attending the seminar were high.
Corrected: The number of students attending the seminar was high.
Explanation: The phrase "the number of" is followed by a singular verb. Here, "number" is the subject, which is singular, so the verb should be "was".

Exercise 7
Incorrect: Each of the players in the team are responsible for their training.
Corrected: Each of the players in the team is responsible for their training.
Explanation: "Each" is always followed by a singular verb, so "is" should be used instead of "are".

Exercise 8
Incorrect: The data suggest that the hypothesis are correct.
Corrected: The data suggest that the hypothesis is correct.
Explanation: "Hypothesis" is a singular noun, so it requires the singular verb "is". "Data" can be plural, so "suggest" is appropriate.

Exercise 9
Incorrect: Everybody in the office need a break.
Corrected: Everybody in the office needs a break.
Explanation: Indefinite pronouns like "everybody" are singular and require singular verbs. Therefore, "needs" is the correct verb form.

Exercise 10
Incorrect: Neither of the solutions works.
Corrected: Neither of the solutions works.
Explanation: This sentence is actually correct. "Neither" is followed by a singular verb, so "works" is appropriate.

……

◆ 解读与反思

　　GPT-4 设计的 20 个主谓一致练习基本覆盖了英语语法中常见的主谓一致错误类型，如单数/复数形式错误（she go、dogs bark）、并列主语（John and Mary is）、集合名词作主语（team have won、committee disagree）、不定代词作主语（everybody need、each of the players are）、主语包含插入语（The teacher along with … were）和特定名词或短语作主语（Mathematics are、The series of events have）等问题。对于大语言模型生成的内容，教师可进一步采取"人工智能后编辑"的方式，使其生成内容更加符合教学要求。

2.3　大语言模型在听力教学中的应用

2.3.1　听力材料改编

◆ 问题与目标

在教学中使用大语言模型将阅读材料改编为听力材料具有多种优势。教师在准备听力材料时，需考虑听力材料的主题、难度等多种因素，这些因素对教师选择听力教学材料具有较多限制。使用大语言模型将阅读材料改编为听力材料，将极大地促进听力教学活动。首先，海量的阅读材料为听力材料的改编提供了丰富的语域和主题，使学生能够接触到多种语言风格和词汇；其次，大语言模型可以根据学生水平定制听力材料，调整听力材料的难度，使教学具有个性；最后，大语言模型可快速进行改编，帮助教师节省大量时间。以下为 GPT-4 针对《中国日报》（*China Daily*）的文章"Chinese Bus Wins Double Busworld Europe Award"所改编的听力材料，改编难度为欧框 B1 水平。

◆ 提示与结果

🔊 提示语（有删减）

Please modify the following reading material into a CEFR-B1-level listening material.

Here is the material:
BRUSSELS – The T15E series bus from the Chinese brand Yutong received a double award from Busworld Europe, both the "Label of Design Excellence" and the "Label of Ecological Excellence" on Saturday at the biennial trade show in Brussels.
Yutong's T15E model, an ultra-luxurious 15-meter-long battery-powered tourist bus, was presented with excellent performance in terms of energy consumption, thanks to its lightweight design and energy-saving technology.
This 26th edition of Busworld Europe, which runs until Oct 12, saw the participation of 526 exhibitors from manufacturers in Europe, Asia, and Africa.
In recent years, Chinese buses have gradually moved towards the development of a "high-end brand" image, and the Busworld expo provides opportunities for Chinese bus makers to enter the global market.

Chinese bus makers were conspicuous at the industry platform. In addition to Yutong, these included BYD, Golden Dragon, and Higer. These Chinese manufacturers showcased their technological innovations and respect for the environment.

BYD, the world's leading manufacturer of new energy vehicles, presented two of its new ebus models at the event.

The Higer presented its latest model, the "Fencer", for the first time, which has been nominated for the Busworld Europe 2023 Vehicle Awards.

Another major Chinese bus manufacturer, Zhongtong Bus, attracted the attention of visitors with its hydrogen fuel model N12.

适用模型：GPT-4

结果（有删减）

Narrator: Let's talk about an exciting event in Brussels. There was a big bus show called Busworld Europe.

Host: Right, and a special Chinese bus won two awards. The bus is called the T15E from Yutong. It won the "Design Excellence" award and the "Ecological Excellence" award. This bus is special because it runs on a battery, is 15 meters long, and uses less energy because of its design.

Narrator: Exactly. This Busworld Europe show happens every two years. This time, 526 companies came to show their buses from places like Europe, Asia, and Africa. Chinese buses are getting more popular. At this show, there were many Chinese companies.

Host: Oh, which other Chinese companies were there?

Narrator: There were many, like BYD, Golden Dragon, and Higer. They all showed their new bus designs and how they care for the environment. BYD, which is a big company, showed two new bus models. Higer showed a new model called the "Fencer". And another Chinese company, Zhongtong Bus, showed a bus that runs on hydrogen.

This version simplifies the material, uses less complex sentence structures, and reduces some of the more specific details to make it suitable for a B1 listening comprehension level.

◆ 解读与反思

大语言模型将阅读材料改编为听力材料，主要做了以下三个方面的修改：（1）在行文上将书面语的语言使用修改为口语的语言使用；（2）将内容形式改编为对话形式，以增加听力材料的吸引力；（3）将改编的听力材料的难度控制在欧框 B1 水平。听力材料的改编保留了原文的主要内容，包括客车展览和中国客车制造商等关键信息。教师可通过修改提示语来控制听力材料的难度，以使生成的结果符合听力教学的需求。对于生成的听力材料，教师可借助语音合成技术、音频处理软件来制作听力内容，或聘请专业人员进行录制。

2.3.2 听力习题设计

◆ 问题与目标

听力题目可分为多种类型，包括信息理解题（提取听力材料的特定信息和细节）、主旨大意题（考查对听力材料的整体理解）和推断题（根据听力材料进行合理推断）等。听力题目的题型包含单项/多项选择题、判断题、填空题、排序题、主观题等。大语言模型可根据相应的要求对输入的听力材料设计题目，帮助教师快速完成听力教学的准备工作。以下为 GPT-4 针对 2.3.1 节中大语言模型改编的听力材料所设计的听力题目，包括两道信息理解单选题、一道信息理解判断题、一道推断题和一道总结主旨大意的主观题。

◆ 提示与结果

> 🔊 **提示语（有删减）**

Please create listening comprehension questions based on the provided listening material, including 2 multiple-choice questions for information understanding, 1 true/false question for information understanding, 1 inference question, and 1 subjective question to summarize the main idea. Please provide answers and explanation.

Here is the material:
Narrator: Let's talk about an exciting event in Brussels. There was a big bus show called Busworld Europe.

Host: Right, and a special Chinese bus won two awards. The bus is called the T15E from Yutong. It won the "Design Excellence" award and the "Ecological Excellence" award. This bus is special because it runs on a battery, is 15 meters long, and uses less energy because of its design.

Narrator: Exactly. This Busworld Europe show happens every two years. This time, 526 companies came to show their buses from places like Europe, Asia, and Africa. Chinese buses are getting more popular. At this show, there were many Chinese companies.

Host: Oh, which other Chinese companies were there?

Narrator: There were many, like BYD, Golden Dragon, and Higer. They all showed their new bus designs and how they care for the environment. BYD, which is a big company, showed two new bus models. Higer showed a new model called the "Fencer". And another Chinese company, Zhongtong Bus, showed a bus that runs on hydrogen.

适用模型：GPT-4

结果（有删减）

Listening Comprehension Questions

Multiple-choice Questions (Information Understanding)

a. What is the name of the bus show mentioned?
A. Busworld Asia B. Busworld Europe C. Busworld Africa D. Bus Expo Europe
Answers: B. Busworld Europe
Explanation: The material mentions "There was a big bus show called Busworld Europe."

b. Why is the T15E from Yutong special?
A. It runs on petrol. B. It uses a lot of energy. C. It is 5 meters long. D. It runs on a battery and uses less energy because of its design.
Answers: D. It runs on a battery and uses less energy because of its design.

Explanation: The material states "This bus is special because it runs on a battery, is 15 meters long, and uses less energy because of its design."

True/False Question (Information Understanding)

True or False: Busworld Europe happens every year.
Answers: False
Explanation: The material says "This Busworld Europe show happens every two years."

Inference Question

From the information given, which of the following can be inferred about Chinese buses?
A. They are not very popular. B. They often win awards. C. They are increasingly focusing on environmental designs. D. They are the only ones displayed at Busworld Europe.
Answers: C. They are increasingly focusing on environmental designs.
Explanation: The material discusses how the T15E won an ecological award, and other companies showed their environmentally-friendly designs, like the bus that runs on hydrogen.

Subjective Question (Summarize the Main Idea)

In a sentence, what is the main idea of this listening material?
Answers: The listening material focuses on Busworld Europe, where Chinese buses, particularly the T15E from Yutong, were showcased and recognized for their innovative and ecological designs.
Explanation: The material revolves around the event Busworld Europe and the increasing popularity and recognition of Chinese buses for their designs and environmental considerations.

◆ 解读与反思

大语言模型针对听力材料所设计的听力题目基本符合提示语的要求,生成的问题涵盖了信息理解题、主旨大意题和推断题三种类型,包括单选题、判断题、主观题等题型。通过修改提示语,教师可以限定题目的难易程度,也可对题目的数量、类型和考查方式进行调整。对于大语言模型所设计的题目,教师可采取"人工智能后编辑"的方式进一步修改,以符合听力教学的要求。

2.3.3 听力音频制作

◆ 问题与目标

在 GPT-4 的 Plugins 模式下,可以安装特定插件,实现将文字材料转换为音频格式的功能。以 QuickVoice 插件为例,GPT-4 能够将文本内容转化为音频文件,并且可以设定语速和说话人的口音。

◆ 提示与结果

🔊 提示语(有删减)

Could you please generate an audio file of the provided text, narrated in a British accent at a speed of approximately 100 words per minute?

Here is the text:
Exciting news for tourists from China's mainland, Hong Kong, and Macao! In mid-December, there's a special chance to see the Hong Kong–Zhuhai–Macao Bridge. It's the world's longest bridge over the sea, and there will be a trial run for tour groups.
Here's what you need to know about the tour:
1. Starting Point: The tour starts at the Zhuhai port.
2. Route: You'll travel to the Blue Dolphin Island, a man-made island on the bridge.
3. Duration: The tour takes about 140 minutes.
4. Views: You'll see three special bridges: the Qingzhou, shaped like a Chinese knot, the dolphin-shaped Jianghai, and the Jiuzhou, which looks like a sail. You might even see the Chinese white dolphin in the sea!
5. Who Can Join: Residents of Hong Kong and Macao with valid permits can join, along with mainland tourists.

6. Rules: Stay with your group and follow the tour path.
7. Travel to Hong Kong and Macao: If you want to go there, you need to go back to Zhuhai port first.
8. Booking: You need to book in advance. Guides who can explain things to you will be there.
9. Safety and Enjoyment: The tours are part of making sure the bridge experience is safe, smooth, and enjoyable.

Since the bridge opened in 2018, 36 million people have visited it. Half of them are from Hong Kong and Macao. Timothy Chui, a travel expert in Hong Kong, says these tours are great for seeing the beauty of the area and learning about the bridge's construction. He hopes there will be similar tours starting from Hong Kong too.

适用模型：GPT-4（使用 QuickVoice 插件）

✅ 结果（有删减）

The audio file of the provided text, narrated in a British accent, has been generated. You can listen to it by clicking on the following link: Listen to the Audio File
（示例音频文件可从配套网站读取）

◆ 解读与反思

　　GPT-4 基于 QuickVoice 插件将文字材料转换为效果较佳的音频材料，且基本满足了每分钟 100 词的速度和英式口音的要求。在超音段音位层面，生成的音频材料对于重音、连读、停顿等的把握基本符合英语母语者的口语习惯。教师可进一步编写提示语，引导大语言模型调整所生成音频的语速，以及说话人的口音、性别等。

2.3.4　听力词表制作[1]

◆ 问题与目标

　　对于学生不熟悉的听力材料主题，提前给出听力生词表——包含材料中出

---

[1] 本案例中的新闻转写来自 CGTN 新闻 "Global Digital Trade Expo: Participants Eye Future Digital Cooperation with China"。

现的术语及语域中的特色高频词汇短语供学生课前预习，可在一定程度上降低听力练习的难度，使课堂节奏更加流畅，并增强学生的成就感。听力生词表还方便学生课后复习，可增进学生对特定主题或语域内容的熟悉程度。

本案例以英语新闻的转写文本为例，展示大语言模型基于新闻自动生成的辅助材料。除包含词汇、术语解释的一般词表外，我们还可进一步追问，要求大语言模型针对特定词汇给出在新闻语境中的常见搭配和相应例句，帮助学生了解词汇在新闻文本中的典型用法。

◆ 提示与结果

🔊 提示语（有删减）

You are an English news tutor, specializing in helping English learners. Your task is to create vocabulary lists from English news transcripts. For a given transcript, please follow these steps:

1. Identify and list uncommon words, along with highly frequent words and phrases that are typical in news reporting.
2. Provide the Chinese translation for each identified English term.
3. Clarify the grammatical usage of each term in English.
4. For named entities (like people, places, organizations) mentioned in the news, provide concise English explanations to clarify their relevance in the context.
5. Upon request, give examples of how a target expression is used, and list its typical collocations (words or phrases that frequently accompany the target expression).

As part of the second Global Digital Trade Expo, delegates from China's Silk Road E-commerce partner countries and international organizations visited the city's leading e-commerce and digital technology companies, such as Alibaba and Hikvision. They told CGTN their views on China's digital development and prospects for future cooperation.

……

适用模型：GPT-4

> **结果（有删减）**

Based on the provided news transcript, here's a vocabulary list with translations, grammatical usages, and explanations for named entities:

1. **Global Digital Trade Expo**
   - Chinese: 全球数字贸易博览会
   - Usage: Noun Phrase
   - Explanation: An international event focused on digital trade and e-commerce.
2. **Delegates**
   - Chinese: 代表
   - Usage: Noun
   - Grammatical Note: Often used in plural form to refer to representatives from various entities.
3. **Silk Road E-commerce**
   - Chinese: 丝绸之路电子商务
   - Usage: Noun Phrase
   - Explanation: Refers to e-commerce initiatives linked to the Silk Road, a historical network of trade routes.
4. **CGTN**
   - Chinese: 中国环球电视网
   - Usage: Acronym
   - Explanation: China Global Television Network, an international English-language news media.
5. **Vice Rector**
   - Chinese: 副校长
   - Usage: Noun
   - Grammatical Note: A title used in educational institutions, typically below a rector or president in hierarchy.
6. **Commercialization**
   - Chinese: 商业化
   - Usage: Noun
   - Grammatical Note: The process of bringing a product or service to market.
7. **Digital Technologies**
   - Chinese: 数字技术
   - Usage: Noun Phrase

- Grammatical Note: Refers to electronic tools, systems, devices, and resources that generate, store, or process data.

  …

🔊 提示语—追问（有删减）

Could you give me some examples and the typical collocations of the word "delegates" in the context of the news?

适用模型：GPT-4

✅ 结果（有删减）

Certainly! In the context of the news transcript, "delegates" refers to representatives, typically from various countries, organizations, or institutions, who are attending a conference, summit, or similar event. Here are some examples and typical collocations for the term "delegates" in this context:

1. **Examples of Usage:**
   - "The conference was attended by delegates from over 30 countries."
   - "Delegates at the summit discussed global trade issues."
   - "Each member state sent two delegates to the annual meeting."
2. **Typical Collocations:**
   - **Send delegates:** Referring to the act of appointing or dispatching representatives.
   Example: "Many nations will send delegates to the international forum."
   - **Meeting with delegates:** The act of holding discussions or negotiations with representatives.
   Example: "The president is meeting with delegates from several African countries."
   - **Delegates representing:** Used to specify whom the delegates are representing.
   Example: "Delegates representing various non-governmental organizations voiced their concerns."
   - **Elected delegates:** Refers to delegates chosen through a voting process.
   Example: "Elected delegates will have the right to vote on the policy changes."
   - **International delegates:** Delegates from different countries around the world.
   Example: "International delegates arrived for the climate summit."

These examples and collocations illustrate the common usage of "delegates" in a context similar to that of your news transcript.

◆ 解读与反思

大语言模型生成的生词材料基本符合要求，此外还提供了目标词在新闻语境中的常见搭配和例句，能够帮助学生熟悉词汇或短语在同类文本中的典型用法。但大模型生成的结果在两个方面仍需人工校验。一是对生词的选择可能与教学需求不符，如原文中 prospect 一词在新闻中其实较常见，但未被大模型选入词表。二是大模型的中文翻译需人工核实。例如案例中 CGTN 被翻译为"中国环球电视网"，但正式名称应为"中国国际电视台"。因此教师应结合教学经验严格把关。

2.4 大语言模型在口语教学中的应用

2.4.1 小组辩论的立论与驳论内容促成

◆ 问题与目标

在"产出导向法"的内容促成环节，教师可引导学生针对特定话题不同视角的多个正反理据表达个人观点，展开辩论，由此形成学生的个人见解（邱琳 2017）。在这个过程中，大语言模型可根据特定话题生成多个正反理据，供教师备课使用或供学生参考学习。以下为 GPT-4 针对"人工智能生成的画作是艺术吗？"这一话题从多个视角生成的正反理据。

◆ 提示与结果

🔊 提示语（有删减）

Please create multiple arguments for and against the topic of "whether paintings generated by artificial intelligence are considered art" from the viewpoints of diverse groups, in order to guide students in constructing arguments and counterarguments.

适用模型：GPT-4

> **结果（有删减）**

Arguments for AI-generated paintings being considered art

1. Technological advancement perspective:
- Argument: AI represents the forefront of human technological advancement. Just as photography was eventually accepted as an art form, AI-generated paintings should also be recognized as a new, legitimate medium of artistic expression.
- Counterargument: Technological innovation does not inherently confer artistic status. The creative intention and human experience behind art cannot be replicated by machines.

2. Postmodern art theory:
- Argument: Postmodernism challenges traditional boundaries and definitions of art. AI-generated paintings, as a new form of expression, fit perfectly into this paradigm, questioning and expanding what art can be.
- Counterargument: Postmodernism still relies on human insight and intention, which AI lacks. Art is not just about breaking boundaries but also about human expression and interpretation.

…

Arguments against AI-generated paintings being considered art

1. Human creativity and intention:
- Argument: Art fundamentally requires human creativity and intention. AI, lacking consciousness and emotional experiences, cannot create art in the true sense.
- Counterargument: Human artists can use AI as a tool to express their creativity, similar to how painters use brushes.

2. Historical and cultural context:
- Argument: Art is deeply rooted in human history and culture, reflecting the human condition. AI-generated paintings, devoid of this context, cannot encapsulate the essence of art.
- Counterargument: All new art forms initially lack historical context but can grow to develop their own cultural significance.

…

◆ 解读与反思

大语言模型较好地就"人工智能生成的画作是艺术吗?"这一话题从多个视角生成了正反理据,包括科学技术进步、后现代艺术、人类创造力、艺术史等角度,从而引导学生从多个视角出发看待问题。大语言模型生成的论据简洁有力,为学生提供了良好的起点,可引导学生进一步深化自己的论证。对于大语言模型生成的结果,教师可采取"人工智能后编辑"的方式进一步修改正反理据以符合口语教学需求,并以此引导学生分为正反两组进行立论与驳论,从而完成内容促成过程。

2.4.2 音节接龙语言促成活动设计

◆ 问题与目标

"音节接龙"是改善语言产出困难的重要方式。在"产出导向法"语言促成理论的指导下,教师可带领学生学习多音节单词的音节知识,切分音节单位,并引导学生在小组内完成续接音节组词的任务,以解决学生在多音节单词发音和拼写方面存在的困难(邱琳 2017)。以下为 GPT-4 生成的与"恶化"意义相关的多音节单词(三个音节及以上)的音节切分。

◆ 提示与结果

🔊 提示语(有删减)

Please generate a list of 10 multisyllabic English words (comprising three more syllables) that are associated with the notion of "deterioration". Additionally, include the syllable divisions for each word.

适用模型:GPT-4

✅ 结果(有删减)

Here is a list of 10 multisyllabic English words related to the concept of "deterioration", along with their syllable divisions:
Degeneration—de-gen-er-a-tion
Decomposition—de-com-po-si-tion

Disintegration—dis-in-te-gra-tion
Destabilization—de-sta-bil-iza-tion
Deterioration—de-te-ri-or-a-tion
Fragmentation—frag-men-ta-tion
Corrosion—cor-ro-sion
Depreciation—de-pre-ci-a-tion
Devastation—dev-as-ta-tion
Disrepair—dis-re-pair
Each of these words relates to a process or state of decline or breakdown, aligning with the concept of "deterioration".

◆ 解读与反思

从生成的结果来看，大语言模型清晰准确地响应了提示语，列出了与"恶化"意义相关的10个多音节英语词汇，并为每个词提供了音节划分。大语言模型对每个词的音节划分都是无误的，并且以连字符连接各音节，格式清晰整齐。教师可根据教学任务调整提示语，来引导大语言模型生成其他符合主题要求的多音节单词，并做音节切分，由此推动学生完成课堂的"音节接龙"任务。

2.4.3 听后续说任务设计

◆ 问题与目标

从"续论"语言习得观的角度出发，"听后续说"作为"续作"的重要操作之一（其余二者为续写、续译），可同外语口语教学有机结合，发挥重要的促学效果。在大语言模型的帮助下，教师可将输入的阅读材料智能地改写为听力材料，并引导学生进行"听后续说"，对听力材料进行创造性的补全和拓展，发挥听力理解与口语产出之间的拉平效应，并引导学生完成听后续说任务的相关讨论，发挥协同效应（张秀芹、王迎丽 2020）。

以下为 GPT-4 辅助听后续说任务的例子。针对输入的英语故事（取自英语爱情故事 *Love Is in the Air*），GPT-4 删除了故事的部分结尾，并改写为欧框 B2

水平的听力材料，引导学生进行听后续说和讨论。该听力材料还通过 GPT-4 的插件 QuickVoice 转换为每分钟 100 词、英式口音的听力音频。

◆ 提示与结果

> 提示语（有删减）

Please create a listening-speaking integrated continuation task suitable for CEFR B2 level learners, based on the provided story. The listening part should be a 200-word text, converted into an audio file with a British English narration at a pace of 100 words per minute. Additionally, include a continuation speaking sample that logically extends the story's theme from the listening segment. Here is the provided story:

Every summer, Penny travels to a family reunion barbeque. Penny is never excited, and this year is no different. She dreads the drive.
She does not like talking to her relatives. And she does not like the smell of hamburgers. (Penny is a vegetarian.)
When Penny arrives, she sees lots of familiar faces. It is July and Uncle Vernon is wearing a sweater. Uncle Vernon is always cold. It's very mysterious.
She sees her cousin Polly. Polly has six children. The youngest one screams. Then the oldest one screams. Polly's children are always screaming.
She sees many of her other cousins in the field playing softball. They play a softball game every year, and it always ends in a big argument. Penny wonders, again, why they never solve it.
Then Penny sees an incredibly handsome man. She stares at him. He catches her staring. He smiles and walks over to her. Penny is very nervous. She is nervous because a handsome man is walking up to her and she is nervous because this handsome man might be her cousin.
The man sticks out his hand and says, "Hi, I'm Paul."
"Hi, I'm Penny," Penny says. "Are we related?"
Paul laughs. "No, we are not related. I am Vernon's nurse. He is sick and needs to keep me close by. But he did not want to miss this barbeque!"
"Oh, thank goodness," Penny says and then blushes. Penny always blushes when she is nervous, embarrassed, or hot, and right now she is all three.
Handsome Paul laughs and says, "Would you like to go get a hamburger with me? They smell delicious."

Penny smiles, "Sure. I love hamburgers!"

适用模型：GPT-4（使用 QuickVoice 插件）

✅ 结果（有删减）

Penny and Paul walk towards the grill, chatting amiably. Penny finds herself enjoying Paul's company, a rare feeling at these family gatherings. As they approach the grill, Penny hesitates. She remembers she's a vegetarian and doesn't eat hamburgers. She wonders how to tell Paul without seeming rude or ungrateful. Just then, Aunt Mabel, known for her culinary experiments, calls out from a nearby table. "Penny, dear, I remembered you're a vegetarian! I made some special veggie burgers just for you!" Penny is relieved and grateful. She introduces Paul to Aunt Mabel, who winks at Penny knowingly.

While eating, Penny and Paul discuss various topics, from their careers to their hobbies. Penny learns that Paul is passionate about environmental conservation, which resonates with her own values. They laugh, share stories, and for the first time, Penny feels genuinely happy at the family reunion. As the afternoon turns to evening, Penny realizes that this year's event is different. She's made a new friend, and perhaps, something more. The thought makes her smile.

The softball game ends without the usual argument, and everyone gathers around for Uncle Vernon's speech. He thanks everyone for coming and makes a special mention of Paul's dedication and care. Penny looks at Paul and feels a warm sense of admiration. As the day ends, Penny finds herself looking forward to the next family reunion, a feeling she never thought she'd have.

The listening part of the task is ready. You can listen to the 200-word text, narrated in British English, by clicking on the link below. This narration is paced at approximately 100 words per minute, suitable for CEFR B2 level learners.

Listen to the Audio
（示例音频文件可从配套网站读取）
Speaking Task (Continuation)

Instructions for Learners:
Now that you have listened to the story about Penny and Paul at the family reunion, imagine you are Penny. Continue the story by describing the rest of your day at the reunion. Focus on your interactions with Paul, your family members, and how your feelings about the family reunion have changed. Use a variety of complex sentences and vocabulary appropriate for the B2 level. Your response should be about 2 minutes long.
……

◆ 解读与反思

GPT-4 较好地基于输入的故事 *Love Is in the Air* 设计了听后续说任务。在设计听力材料时，GPT-4 智能地去掉了结尾部分，并将原文改写为欧框 B2 水平的 200 词阅读材料；又使用了插件 QuickVoice 将听力材料转变为听力音频。此外，GPT-4 设计的听后续说任务引导学生设想自己是故事中的主角 Penny，用两分钟的时间描述在家庭聚会的余下时间里所发生的故事，包括与 Paul 的交流以及同其他家庭成员之间的互动。该任务生动有趣，能够有效引导学生延续听力部分的故事，创造性地进行口语产出，从而提高学生的口语表达能力。

2.4.4 "视后续说"任务设计

◆ 问题与目标

从"续论"语言习得观出发，"续作"作为促学语言方法可将语言理解与语言产出紧密联系。在语言理解方面，续作的输入方式主要有视、听、读三种，但既往研究中基于视（看图像）所开展的续说、续写、续译研究较少，部分原因在于图像、视频等配套资源相对缺乏。本节旨在说明使用大语言模型的图片生成功能为视后续说任务配图的可行性。以下为通义万相以 Robert Southey 的童话故事 *Goldilocks and the Three Bears* 的节选为提示语生成配图的例子。教师可使用特定提示语生成图片，引导学生开展视后续说的教学任务。

◆ 提示与结果

🔊 提示语（有删减）

图片 1：Once upon a time a girl named Goldilocks lived in a house at the edge of the woods. In those days curls of hair were called "locks". She was "Goldilocks" because golden hair ran down her head and shoulders. One morning Goldilocks was out for a walk when she came across a beautiful bird.

图片 2：But a house was not far away. "I wonder who lives there," she thought, "so deep into the woods." She went up and knocked on the door. No answer. She knocked again. Still no answer. Goldilocks knocked a third time and the door opened.

适用模型：通义万相

✅ 结果（有删减）

图片 1：

图片 2:

◆ 解读与反思

  整体来讲，大模型很好地捕捉了提示语描述的场景，细节较为丰富，色彩和构图也较为出色。如第一张图片中，小女孩金色的卷发确实与 Goldilocks 这个名字相匹配；小女孩与鸟儿的互动也与 came across a beautiful bird 相吻合。在第二张图片中，阳光从树间倾泻而下，为画面带来了一种安详和温暖的氛围，小女孩沿着小路向前走，展现了她好奇的性格。两张图片在风格上也存在连贯性，为"视听后续说"任务的开展提供了支持。但值得注意的是，提示语中的一些细节未能完全展现出来，如图片 1 提示语中的 a house at the edge of the woods、图片 2 提示语中的 knocked on the door 等，且图片 2 中生成的房子也不完整。

  在实际操作中，提示语不宜过长，否则大模型将遗漏某些细节。如生成结果不匹配，可使大模型重复生成，直至满意。使用大模型为现有听力、阅读材料生成配图，可大大弥补"视听后续说"教学任务中图像资源匮乏的问题，将有力促进该教学活动的开展。

2.4.5　英语演讲讲稿修改和即兴问答设计

◆ 问题与目标

英语演讲不仅是高端外语人才的必备技能，也体现了外语人才在外语能力、思辨能力、交际能力、创新能力等方面的综合素质。大语言模型可辅助修改英语演讲稿，并设计即兴问答题目，帮助英语演讲比赛选手更好地备赛。以下内容展示了 GPT-4 从语法、不自然搭配、与演讲语体不相符的表达三个方面对定题演讲稿的修改，以及基于该演讲稿所设计的即兴问答题目，该定题演讲的题目为"技术对教育的影响"（The Impact of Technology on Education）。

◆ 提示与结果

🔊 提示语（有删减）

Please revise the provided speech manuscript titled "The Impact of Technology on Education" using in-text annotations. Focus on correcting any grammatical errors, unnatural collocations, and expressions that are not suitable for a speech format. Additionally, create 5 impromptu question-and-answer items based on the provided speech manuscript.

In recent decades, the impact of technology on education has been profound and multifaceted. The advancements in technology have not only transformed the methods and means of education but also expanded its reach and efficiency. Firstly, technology has revolutionized the accessibility of education. Online courses and e-learning platforms have made education accessible to a wider audience, breaking geographical barriers. Students from remote areas can now access quality education from top institutions around the world. This democratization of education has been a significant step towards educational equity. Moreover, technology has introduced a variety of learning tools that cater to different learning styles. Interactive software, educational apps, and multimedia content like videos and podcasts offer diverse ways of learning, making education more engaging and effective. These tools also provide personalized learning experiences, allowing students to learn at their own pace and according to their own learning style.

……

适用模型：GPT-4

结果（有删减）

Revised Speech Manuscript: "The Impact of Technology on Education"

[Original Text] In recent decades, the impact of technology on education has been profound and multifaceted.
[Revised] Over the past few decades, technology's role in education has evolved remarkably, impacting it in numerous and complex ways.

[Original Text] The advancements in technology have not only transformed the methods and means of education but also expanded its reach and efficiency.
[Revised] Technological advancements have revolutionized educational methods and tools, greatly expanding their reach and efficiency.

[Original Text] Firstly, technology has revolutionized the accessibility of education.
[Revised] First, let's consider how technology has dramatically increased education's accessibility.

[Original Text] Students from remote areas can now access quality education from top institutions around the world.
[Revised] Students in remote areas now have the opportunity to access high-quality education from leading global institutions.

[Original Text] This democratization of education has been a significant step towards educational equity.
[Revised] This democratization represents a significant stride toward educational equality.
…

Impromptu Q&A Based on the Speech

1. Q: How has technology improved the accessibility of education?
   A: Technology has made education more accessible by enabling online courses and e-learning platforms, allowing students, even in remote areas, to access quality education from top institutions worldwide.

2. Q: Can you give examples of how technology caters to different learning styles?
   A: Yes, technology caters to various learning styles through interactive software, educational apps, and multimedia content like videos and podcasts, which make learning more engaging and personalized.
3. Q: How has technology facilitated collaboration and communication in education?
   A: Technology has fostered collaboration and communication through platforms which help students and teachers interact, share resources, and work on projects together, regardless of their physical locations.
   ……

◆ 解读与反思

在修改演讲稿的任务中，GPT-4 将较为书面化的语句改为更适合口语的措辞，使其更易于理解；GPT-4 也对原文进行了重组，以确保内容适合口头演讲；最后，GPT-4 对讲稿中的某些关键点进行了强调，使其更具吸引力。对于这一定题演讲，GPT-4 从技术在教育中的作用、个性化学习以及技术依赖带来的挑战等方面设计了即兴问答题目。教师可采取"人工智能后编辑"的方式，对大语言模型修改的讲稿和设计的问答进一步修改，使其符合教学要求。

2.5 大语言模型在阅读教学中的应用

2.5.1 阅读材料编制

◆ 问题与目标

在以往的阅读教学中，教师常常遇到阅读材料难度与学生的英语水平不相匹配的问题。阅读材料过于简单会导致达不到外语学习效果，阅读材料过于困难则会挫败学生的学习热情。因此，教师需要花费较多时间和精力来筛选和准备合适的阅读教学材料，例如修改过难或过易的语言表达以符合教学要求。在大语言模型的帮助下，教师可进行定制化的阅读材料设计。以下为 GPT-4 基于《中国日报》文章 "Domestic Road Network Grew by over 1 Million Kilometers in Past Decade" 所设计的欧框 B1 水平阅读材料。

◆ 提示与结果

🔊 提示语（有删减）

Please revise the provided text to match a CEFR-B1 proficiency level, suitable for intermediate language learners. Use in-text annotations to highlight the modifications made.

The text for revision is as follows:
China's expressway network had extended to 177,000 kilometers by the end of last year, ranking top in the world, according to the Ministry of Transport on Thursday.
By the end of last year, China's road network reached 5.35 million km, an increase of 1.12 million km over the past decade, said Guo Sheng, deputy head of the ministry's highway bureau.
"In the past decade, historic achievements were made in highway construction and development in China," he said.
During the past 10 years, major national expressways linking big cities have been expanded and upgraded, such as the Beijing–Shanghai, Beijing–Hong Kong–Macao, Shenyang–Haikou and Shanghai–Kunming.
National expressways with more than six lanes have increased by 18,400 km over the 10 years, he noted.
A great number of major infrastructure projects have been built and created new bench marks for expressway infrastructure, he said.
The Hong Kong–Zhuhai–Macao Bridge, one of the world's most challenging infrastructure projects, has been put into operation.
More than 10 bridges spanning the Yangtze River have been built and opened to traffic. The Heihe Bridge, built in cooperation with Russia, has been built and opened to traffic.
The expressway linking Beijing and the Xinjiang Uygur autonomous region—with a total length of over 2,500 km—was completed.
Mega projects such as the Hong Kong–Zhuhai–Macao Bridge and the Shenzhen–Zhongshan link have tackled new technologies and achieved breakthroughs, such as being able to construct and place large-sized undersea tubes.
According to the ministry, there were 5.34 billion cross-regional passenger trips last month, a year-on-year increase of 50.2 percent.

适用模型：GPT-4

> 结果（有删减）

Here's the revised text suitable for CEFR-B1 language learners, with annotations highlighting the changes made:
1. **China's expressway network** [Simplified to "China's big roads"] **had extended to 177,000 kilometers** [Changed to "grew to 177,000 kilometers"] **by the end of last year, ranking top in the world, according to the Ministry of Transport on Thursday.**
   - China's big roads grew to 177,000 kilometers by the end of last year. This is the longest in the world, the Ministry of Transport said on Thursday.
2. **By the end of last year, China's road network reached 5.35 million km, an increase of 1.12 million km over the past decade, said Guo Sheng, deputy head of the ministry's highway bureau.**
   - By the end of last year, China's roads were 5.35 million km long. This is 1.12 million km more than ten years ago, said Guo Sheng, a leader in the road department.
3. **"In the past decade, historic achievements were made in highway construction and development in China," he said.**
   - "In the last ten years, we did great things in building and improving roads in China," he said.
4. **During the past 10 years, major national expressways linking big cities have been expanded and upgraded, such as the Beijing-Shanghai, Beijing-Hong Kong-Macao, Shenyang-Haikou and Shanghai-Kunming.**
   - In the last ten years, big roads connecting big cities like Beijing-Shanghai and others were made bigger and better.
5. **National expressways with more than six lanes have increased by 18,400 km over the 10 years, he noted.**
   - Roads with more than six lanes grew by 18,400 km in ten years, he added.
6. **A great number of major infrastructure projects have been built and created new bench marks for expressway infrastructure, he said.**
   - Many big projects have been built, setting new standards for roads, he said.
7. **The Hong Kong-Zhuhai-Macao Bridge, one of the world's most challenging infrastructure projects, has been put into operation.**
   - The Hong Kong-Zhuhai-Macao Bridge, a very difficult project, is now being used.

8. More than 10 bridges spanning the Yangtze River have been built and opened to traffic. The Heihe Bridge, built in cooperation with Russia, has been built and opened to traffic.
   - Over 10 bridges over the Yangtze River are now open. The Heihe Bridge, made with Russia, is also open.
9. The expressway linking Beijing and the Xinjiang Uygur autonomous region—with a total length of over 2,500 km—was completed.
   - The road from Beijing to the Xinjiang Uygur area, more than 2,500 km long, is finished.
10. Mega projects such as the Hong Kong–Zhuhai–Macao Bridge and the Shenzhen–Zhongshan link have tackled new technologies and achieved breakthroughs, such as being able to construct and place large-sized undersea tubes.
    - Big projects like the Hong Kong–Zhuhai–Macao Bridge used new technologies and made big achievements, like building big tubes under the sea.
    ……

◆ 解读与反思

GPT-4 为将这篇文章转为适合欧框 B1 水平学生的阅读材料，进行了多方面的修改。首先是简化短语，如将 historic achievements 修改为 great things；第二是缩短句长；第三，降低句法复杂度，如将 has been put into operation 修改为 is now being used，并简化时态和语态；第四，具体化描述，如将 achieved breakthroughs 修改为 made big achievements，使其更具体，易于理解。但同时也出现了一些错误的修改，特别是术语的修改，如"中国高速公路网络"由 China's expressway network 修改成了 China's big roads。因此，教师应采取"人工智能后编辑"的方式，对大语言模型生成的内容加以判别，进一步修改以使其符合阅读教学的要求。

2.5.2　阅读习题设计

◆ 问题与目标

大语言模型可根据阅读理解的文章定制题目，助力教师迅速锁定核心议题，从而节省时间。以下是 GPT-4 设计的五道阅读理解选择题，包括细节理解题、推理判断题、观点态度题和主旨大意题等四种类型,旨在全面评估学生对文章的理解。

◆ 提示与结果

> 🔊 提示语(有删减)

Please create five multiple-choice reading comprehension questions based on the given text. These should include Detail, Attitude, Inference, and Main Idea Questions. Provide the correct answers with explanations for each question.

Here is the text:
China is set to enhance its Belt and Road Initiative (BRI) over the next decade, aiming to create a more open, inclusive, and interconnected world. This plan involves integrating green and digital advancements, as outlined in a recent report. The BRI, known as the largest platform for international cooperation, has greatly contributed to global development. Moving forward, China's focus will shift to multilateralism, free trade, and especially green and digital development. Key areas of focus include policy harmonization, infrastructure development, trade facilitation, financial integration, and people-to-people connections. Emerging fields will also be explored. The "Vision and Actions for High-Quality Belt and Road Cooperation" document emphasizes aligning international standards and expanding online trade and investments. Since its inception, the BRI has involved over 150 countries and 30 international organizations. Trade and investments have flourished, with $19.1 trillion in trade and over $380 billion in investments between China and BRI countries from 2013 to 2022. Despite global challenges, optimism remains for the BRI's future, driven by the need for cooperation in green and digital sectors and China's robust industrial system. The initiative is expected to bring mutual gains globally, as noted by experts like Victor Stolzenburg from the World Trade Organization. China anticipates further opening its economy, especially in services, to enhance trade with BRI countries. Financial cooperation is also a focus, with suggestions like a currency reserve mechanism among BRICS countries to promote economic stability. In conclusion, China's BRI is entering a new phase, focusing on sustainable and digital growth. The initiative aims to deepen international partnerships and promote a prosperous future.

适用模型:GPT-4

✓ 结果（有删减）

1. Detail Question: What is the total trade value between China and BRI countries from 2013 to 2022?
   - A) $19.1 billion - B) $380 billion - C) $19.1 trillion - D) $380 million
   - Correct Answer: C) $19.1 trillion
   - Explanation: The text states that trade between China and BRI countries amounted to "$19.1 trillion in trade" from 2013 to 2022.
2. Attitude Question: What tone does the author use when discussing the future of the BRI?
   - A) Skeptical - B) Optimistic - C) Indifferent - D) Critical
   - Correct Answer: B) Optimistic
   - Explanation: The text conveys an optimistic tone, particularly when mentioning "optimism remains for the BRI's future" and the expectation of mutual gains globally.
3. Inference Question: Based on the text, what can be inferred about the role of green and digital sectors in the BRI's future?
   - A) They are of minor importance. - B) They are central to the initiative's new phase. - C) They will be gradually phased out. - D) They have no significant impact.
   - Correct Answer: B) They are central to the initiative's new phase.
   - Explanation: The text emphasizes a shift in focus to green and digital development, indicating these sectors are central to the BRI's new direction.
4. Main Idea Question: What is the main objective of China's Belt and Road Initiative (BRI) as outlined in the text?
   - A) To increase China's political influence globally. - B) To create a more interconnected and inclusive world focusing on sustainable and digital growth.
   - C) To solely enhance China's trade and investment opportunities. - D) To compete with other international cooperation platforms.
   - Correct Answer: B) To create a more interconnected and inclusive world focusing on sustainable and digital growth.
   - Explanation: The text highlights the BRI's aim to create a "more open, inclusive, and interconnected world" with a focus on green and digital advancements.
5. Detail Question: What specific suggestion is mentioned for financial cooperation among BRICS countries?
   - A) Increasing trade tariffs - B) Developing a currency reserve mechanism
   - C) Establishing a new international bank - D) Reducing financial integration

- Correct Answer: B) Developing a currency reserve mechanism
- Explanation: The text mentions the suggestion of "a currency reserve mechanism among BRICS countries" to promote economic stability.

◆ 解读与反思

GPT-4 针对文章设计了细节理解题、推理判断题、观点态度题和主旨大意题四个类型的阅读理解题目，并给出答案及详细解释。设计的每个问题都直接关联文本内容，确保了题目的针对性和相关性。对于题目的难度，教师可进一步修改提示语，引导大语言模型以欧框指定水平生成题目。

2.5.3 交互阅读游戏设计

◆ 问题与目标

常规阅读材料常有难度与学习者水平不匹配、主题与学习者兴趣不匹配等问题。单纯的文本输入也有可能降低阅读的趣味性，致使学习者难以坚持。大语言模型的文本生成能力可为学习者提供一种新的阅读形式，即开放式、交互式的阅读游戏，作为传统阅读活动的补充。

本案例使用 GPTs 定制交互阅读聊天机器人。辅导者或读者本人可以设定文本风格、故事背景、主要人物等，使阅读材料更符合个人兴趣和语言水平。阅读过程中，读者可以通过简单的回复决定主人公的行动，从而影响剧情走向。这样的阅读体验沉浸感强，趣味性高，能够鼓励学习者在情境中学习词汇等语言知识，并通过交互获得使用外语的参与感和成就感。

◆ 提示与结果

🔊 使用 GPT 助手调整 GPTs。模型描述如下：

Adventure Linguist serves as both an English tutor and Dungeon Master within a fantasy adventure game. It's designed to help users improve their English skills while engaging in an interactive storytelling experience. The GPT will now include explanations for difficult words and phrases at the end of its messages, making it easier for English learners to understand and learn new vocabulary. Additionally, the GPT will provide more suggestions for the player's actions, offering a variety of options to enrich the interactive gameplay experience. This approach enhances language learning by actively involving users in decision-making and story development, while ensuring that the narrative remains accessible and educational. The focus on creativity and avoidance of cliché plots remains, with the GPT continuing to request keywords or themes from users to tailor unique and imaginative stories.

适用模型：GPT-4

✅ 提示与结果（有删减）

User: Hello.
GPT-4：Greetings, traveler! Welcome to a world of fantasy and language learning. I am the Adventure Linguist, here to guide you through enchanting tales while helping you improve your English. Before we embark on our journey, do you have any specific themes or keywords you'd like to explore in our adventure? Also, feel free to ask if you encounter any English words or phrases that are unfamiliar to you. Let's create a story that's both thrilling and educational!

User: Female protagonist, harmony between man and nature, revenge.
GPT-4：(Story begins) Splendid choices! Let's weave a tale about a female protagonist who seeks to restore harmony between humankind and nature, all while grappling with her own quest for revenge. Here's the beginning of our story: In the lush, verdant realm of Eldoria, where the trees whispered ancient secrets and rivers sang melodies of old, there lived a young woman named Aeliana. […] In a tragic turn, Aeliana's family fell victim to the lord's ruthless actions … As Aeliana, you face a critical decision at this crossroads of your journey: 1. Seek Allies in the Enchanted Forest: Venture into the heart of the forest to seek allies among the mystical creatures dwelling there. They might offer wisdom and aid in your quest for harmony and revenge. 2. Confront the Greedy Lord: Plan a direct confrontation with the lord, using your knowledge of the land and its secrets to outwit him. 3. Gather Knowledge and Proof: Embark on a journey to gather evidence of the lord's misdeeds, intending to bring him to justice before the king. […]

User: I will go on a journey to consult the wise witch living on the top of a snow mountain for her knowledge. On my way, I will gather evidence of the lord's misdeeds in every village I visit while I help heal the villagers who are sick. I have a feeling that this will not be an easy trip.

GPT-4: (story progresses) A wise and compassionate choice, Aeliana. Your journey to the snow-capped mountain to seek the wisdom of the witch will be both challenging and enlightening. Along the way, your efforts to gather evidence and aid those in need will undoubtedly shape the destiny of Eldoria. The Journey Begins: As you set out from your village, the path ahead is fraught with challenges. The mountain looms in the distance, its peak shrouded in eternal snow. You travel from village to village, each affected in its own way by the lord's greed. In each place, you listen to the villagers' stories, collect evidence of the lord's misdeeds, and use your botanical knowledge to concoct remedies for the sick. Unexpected Encounters: Your journey is not without surprises. You encounter a variety of characters: a group of traveling minstrels with tales of distant lands, a band of rebels plotting against the lord, and even a mysterious traveler with knowledge of secret paths through the wilderness. Each encounter offers you opportunities to learn and grow, not just as a healer and avenger, but as a person … Now, Aeliana, you stand before the witch's door. How do you wish to proceed? Do you: 1. Present the Evidence: Show the witch the evidence of the lord's misdeeds and ask for her advice on bringing him to justice. 2. Seek Knowledge about Harmony: Inquire about ways to restore harmony between mankind and nature, putting your quest for revenge aside for the moment. […]

◆ 解读与反思

定制大模型生成的故事流畅自然，符合西方奇幻作品的常规风格。但从当前案例的生成结果来看，大模型对故事的描述一般比较笼统，情节发展较快，故事角色也有脸谱化之嫌。一方面，进一步精调提示词可以改善这些不足；另一方面，读者在与大模型互动时也需要充分想象故事情境，这对于学习者的能力也是一定的考验。

由于篇幅限制，案例中没有展示每条信息最后对文本中较难词汇或短语的解释。这一功能能够降低阅读难度，有助于学习者基于情境理解、掌握词汇。

2.5.4 识别争议焦点的观点促成活动设计

◆ 问题与目标

在"产出导向法"观点促成理论所指导的阅读教学中,教师可选取特定阅读材料,引导学生通过句子、段落、篇章阅读来识别争议焦点,由此促进学生进行正反辩论,形成个人观点(张文娟 2017)。以下为 GPT-4 根据话题"人工智能是否会取代人类工作?"生成的一篇有争议性的阅读材料,且 GPT-4 在该材料的句子、段落、篇章层级标注出了争论焦点。

◆ 提示与结果

🔊 提示语(有删减)

Please generate a controversial reading material based on the topic "Will artificial intelligence replace human jobs?" and highlight the points of contention at the sentence, paragraph, and discourse levels within the reading to guide students in engaging in debate.

适用模型:GPT-4

✅ 结果(有删减)

Reading Material: Will Artificial Intelligence Replace Human Jobs?

Introduction
As we venture further into the 21st century, a lingering question remains ever pertinent: will artificial intelligence (AI) replace human jobs? While some champion the advancement of technology as a pathway to easier and more efficient work processes, others fear the impending "robot takeover" and the subsequent unemployment surge it could potentially usher.

Contention at sentence level: The phrase "robot takeover" might be considered inflammatory or fear-mongering, implying a dystopian future which is a highly debatable conception.

Body Paragraph 1: The Case for AI
Proponents of AI argue that technology will foster unprecedented innovation and efficiency in the workplace. Machines do not tire, call in sick, or harbor

personal grievances, allowing for a work output that is both consistent and reliable. Furthermore, automation can handle dangerous tasks, thereby reducing workplace injuries and fatalities. The healthcare industry, for instance, has seen dramatic improvements in diagnostics and patient care thanks to AI applications.

Contention at paragraph level: This paragraph can be seen as overly optimistic, portraying AI as a panacea for workplace issues without addressing potential drawbacks such as job displacement, privacy concerns, and the loss of human touch in various professions.

Body Paragraph 2: The Job Displacement Crisis
Critics warn of a dystopian future where AI displaces workers en masse, leading to economic turmoil and unprecedented levels of unemployment. They argue that while machines can carry out tasks with superhuman efficiency, the widespread adoption of AI could mean fewer opportunities for humans, potentially exacerbating inequalities and fostering social unrest.

Contention at paragraph level: The paragraph paints a grim future without sufficiently acknowledging the potential for new job creation and economic growth arising from AI advancements. The use of the word "dystopian" could be seen as overly dramatic and not grounded in evidence.

……

◆ 解读与反思

　　大语言模型就"人工智能是否会取代人类工作？"这一话题生成了一篇包含多个争议点的阅读材料，以引导学生就此观点展开讨论。从生成的内容来讲，该材料对话题做了较为全面的探讨，从支持、反对、中立的观点出发，内容覆盖得较为全面，并在各个层面突出了争议点，如在句子层面使用robot takeover这样的激进表达引发讨论，在段落层级提出较为绝对的观点来创造潜在的情感争议点。大语言模型对材料中的争议点进行了标注，教师可对照采取"人工智能后编辑"的方式去决定是否采纳每个争议点。总体来说，该材料的生成质量较高。为服务于"产出导向法"观点促成中"识别争议焦点"的教学设计，教师应选取包含一定争议焦点的阅读材料引导学生辩论，但受限于文体风格，该类型材料不便寻找。而大语言模型可针对特定主题、特定焦点有针对性地生成阅读材料，来弥补这方面资源的不足。

2.5.5　阅读教学思维导图绘制

◆ 问题与目标

在阅读教学中，使用大语言模型提取阅读材料的核心要素，如主题句、重要事实、证据等，并使用思维导图等工具进行可视化处理，不仅有助于学生更清晰地把握阅读材料的结构和核心思想，增强他们的理解，还可帮助学生厘清文本中的不同概念，培养思辨能力和分析能力。以下为 GPT-4 在 Plugins 模式下，使用 Diagrams: Show me 插件基于阅读材料生成思维导图的例子。

◆ 提示与结果

> 🔊 提示语（有删减）

Could you create a mindmap based on the provided text?

Here is the text:
Exploring the Interface: The Evolution and Impact of Human-Computer Interaction
Human-computer interaction (HCI) is a multidisciplinary field that focuses on the design of computer technology and the interaction between humans and computers. Initially concerned with computers, HCI has expanded to encompass almost all forms of information technology design.
The field began to take shape in the 1980s with the rise of personal computing. The primary focus during this period was on making computers accessible and usable for individuals without formal training in computer science. This era saw the development of graphical user interfaces (GUIs), which marked a significant shift from text-based command-line interfaces to interactions based on graphical icons and visual indicators.
……

适用模型：GPT-4（使用 Diagrams: Show me 插件）

✓ 结果（有删减）

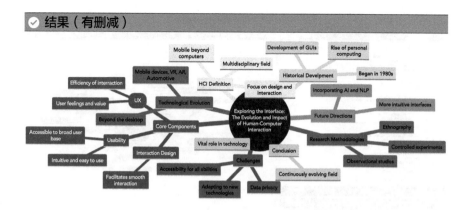

Here is the mindmap diagram illustrating "Exploring the Interface: The Evolution and Impact of Human-Computer Interaction": View fullscreen diagramEdit diagram online. Edit with Miro using drag and drop with a free-forever account. Used mermaid, mindmap. For improvement ideas type "show ideas". Interested in seeing different diagram styles? Here are a few combinations you can try: 1. Mindmap diagram with Markmap 2. Mindmap diagram with PlantUML 3. Sequence diagram with PlantUML

◆ 解读与反思

　　GPT-4 基于 Diagrams: Show me 插件提取了文章 "Exploring the Interface: The Evolution and Impact of Human-Computer Interaction" 中涉及的多个关键要素，如定义、历史发展、核心组成部分、技术演进、面临的挑战等，并以思维导图的形式直观地展示了各主题之间的紧密联系。对于生成的思维导图，还可进一步进入其修改界面，调整内容、位置等。

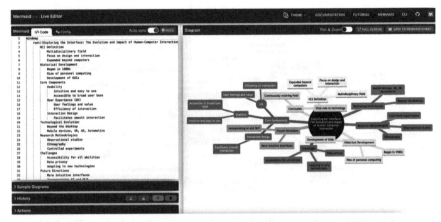

图 2.1　基于 GPT-4 模型生成的思维导图结果及编辑界面

2.5.6　阅读语篇主题词云图绘制

◆ 问题与目标

词云图是一种重要的可视化工具，用于展示文本数据中词汇的频率分布或权重。在词云图中，词汇的频率越高或权重越大，那么该词汇对应的字号就越大。词云图可有效辅助阅读教学任务的开展，它可以帮助学生快速识别阅读材料中的关键词汇和主题，从而引导学生抓住文本的核心概念；在视觉上词云图通过不同字号、不同颜色来引起学生的注意，从而激发学生提出问题并进行探讨。GPT-4 在 Plugins 模式下可引入 WordCloud 插件，来快速实现词云图的制作，并可设置指令来剔除"停用词"（stop word，指在文本处理中通常被过滤或忽略的常见、无实际语义价值的词汇，如冠词）。以下为 GPT-4 在 WordCloud 插件下对《新标准大学英语（第二版）综合教程 4》的课文"If You Ask Me"所生成的词云图，并剔除了词云图中的停用词。

◆ 提示与结果

🔊 提示语（有删减）

Based on the provided text, please generate a word cloud. Please exclude the stop words from the word cloud.

Here is the text:

If you ask me, real life is not all it's cracked up to be. Twelve years at school and three years at university, teachers banging on about opportunities in the big wide world beyond our sheltered life as students, and what do I find?

Try as I might to stay cheerful, all I ever get is hassle, sometimes with people (especially boys—when will they grow up?), but mostly with money. It's just so expensive out here! Everyone wants a slice off you. The Inland Revenue wants to deduct income tax, the bank manager wants repayments on my student loan, the landlord wants the rent, gas, water and electricity, and my mobile bills keep coming in, and all that's before I've had anything to eat. And then some bright spark calls me out of the blue, asking if I'm interested in buying a pension. At this rate, I won't even last till the end of the year, let alone till I'm 60.

……

适用模型：GPT-4（使用 WordCloud 插件）

✓ 结果（有删减）

Alright, I will generate a word cloud for the provided text. Before doing so, I will remove common stop words (e.g., "and", "the", "is", "in", etc.) to ensure the word cloud highlights the most significant and unique words from the text. Here's the word cloud generated from the provided text:

◆ 解读与反思

在大语言模型生成的词云图中，money（钱）、people（人）、job（工作）、year（年份）等在原文中出现频率较高的词汇以较大字号、不同颜色呈现，教师可由此引导学生关注这些词汇对推动故事情节发展和主题的重要意义。与此同时，大语言模型将 and、the、is、in 等停用词去除，以消除这些无实际语义价值的词对词云图的整体影响。教师可通过大语言模型实现阅读材料的可视化，通过词云图来引导学生关注材料中的关键主题，由此推进阅读教学的开展。

2.6 大语言模型在写作教学中的应用

2.6.1 不同语域同题文本生成[1]

◆ 问题与目标

在英语学习过程中，当词汇积累达到一定规模，并掌握基本语法规则后，学习者便进入高级阶段。此时会出现一系列不易克服的学习难题，学生写作中的语体或语域意识（register awareness）就是其中之一，这也是教学组织上的难点。

以学生英语作文中的这一段为例：

The advantages of two-day weekends are too much to be listed. First of all, after busy weekdays, we can relax ourselves, and then we'll be fresh up. Secondly, we can use the two-day weekends to learn something we are interested in. Finally, we can learn the technique which is corresponding to our jobs.（节选自桂诗春、杨惠中 2003）

其中，第一人称代词、情态动态、缩略用法属于偏口语化的表述，而 corresponding 又是个比较正式的词语。

为加强学生对正式、非正式英语的词句风格差异的认识，我们可使用大语言模型就同一主题生成两个语篇，用以对照。

---

[1] 本案例思路受到文秋芳教授启发。

◆ 提示与结果

🔊 提示语

Could you write two 100-word essays on the topic "The Pros and Cons of Artificial Intelligence"? One should be in a formal style, and the other in an informal style.

适用模型：GPT-3.5、GPT-4、Gemini、Claude 2、POE、文心一言、讯飞星火、通义千问

✓ 结果

Formal Style:
Artificial Intelligence (AI) is a rapidly evolving technology that has manifold implications. On the one hand, AI streamlines operations, boosts productivity, and augments human capabilities by automating complex tasks. It holds the potential to revolutionize sectors like healthcare, finance, and transportation. Conversely, AI presents challenges such as job displacement and ethical dilemmas. Moreover, AI's decision-making processes can be opaque, leading to concerns about accountability and bias. Therefore, while AI promises unprecedented advancements, it necessitates careful governance to ensure equitable benefits and mitigate potential pitfalls.

Informal Style:
So, AI's pretty cool, right? It makes our lives easier by doing the hard stuff for us and could totally change how we do things in places like hospitals and banks. But, there's a downside too. Some folks might lose their jobs because a robot can do it, and sometimes these AI systems can be kinda sneaky about how they make decisions, which isn't always fair. We gotta make sure we keep an eye on it, so the cool benefits don't get overshadowed by the not-so-great stuff.

◆ 解读与反思

　　我们可将这两段大语言模型生成的语篇融入课堂教学。比如，可在讲解作文时安排学生分组讨论，总结概括两篇作文的异同。同时再对照个人所写作文的情况，反思作文语域上的问题。

　　当然，教师也可以通过"Could you compare the lexical and grammatical differences between the formal and informal essays?"这样的提示语让大语言模型

指出两篇文章在风格上的差异，还可以用语域分析工具直观展示文章中口语和书面语的特征，具体可参见许家金（2019：96-99）的研究。

我们也可以让大语言模型生成一篇正式和非正式风格混杂的同题作文，让学生通过讨论指出其中的口语化表达。

2.6.2　特定难度主题相关词句推荐

◆ 问题与目标

在基于"产出导向法"的大学英语写作教学中，教师应选取与写作任务、写作目标相契合的语言目标（如与主题相关的常用词与词块、与议论文写作相关的表达）等供教学使用（张文娟 2017）。以下为 GPT-4 基于主题"如何看待大学生创业？"为欧框 B2 水平学习者生成的语言目标，其中包含 30 个与主题相关的常用词与词块，以及与议论文写作相关的 20 种表达。

◆ 提示与结果

🔊 提示语（有删减）

Please create a two-part list for CEFR B2 level English learners. Part 1: List 30 common words and phrases related to the topic of college students starting their own businesses. This should include terminology about entrepreneurship, startup challenges, and business planning. Part 2: List 20 expressions specifically beneficial for argumentative writing. These expressions should help in presenting opinions, making contrasts, providing examples, and concluding arguments. Ensure all terms and expressions are suitable for CEFR B2 level learners.

适用模型：GPT-4

✅ 结果（有删减）

Here's a two-part list tailored for CEFR B2 level English learners:

Part 1: Common words and phrases for college students starting their own businesses
1. Entrepreneurship - The activity of setting up a business.
2. Startup - A newly established business.

3. Funding - Financial support for a business.
4. Revenue - Income generated from business activities.
5. Branding - The process of creating a unique image for a product.
……
Part 2: Expressions for argumentative writing
1. Furthermore - Adding more information.
2. Consequently - Showing the result of something.
3. On the other hand - Presenting an opposing viewpoint.
4. According to - Citing a source.
5. As a result - Showing a consequence.
……

◆ 解读与反思

大语言模型根据提示语较好地生成了与主题相关的关键词语、词汇，以及与议论文写作有关的常用表达。大语言模型所生成的常用词和词块皆与"如何看待大学生创业？"这一主题密切相关；议论文常用表达都能有效地帮助学生更清晰、更有逻辑地表达自己的观点。教师可基于写作主题、写作类型变换提示语，来引导大语言模型生成符合教学任务要求的语言目标。

2.6.3　典型写作交际场景设计

◆ 问题与目标

呈现交际场景是"产出导向法"驱动环节的第一步。在场景设计时，第一应注重交际真实性，第二应聚焦场景要素（话题、目的、身份和场合）及各要素之间的相互关系（文秋芳、孙曙光 2020）。场景要素主要包括以下四个方面：（1）话题（信息/议题、问题/困难）；（2）身份（产出者身份、受众身份）；（3）目的（说明/解释信息、劝说/论证观点）；（4）场合（正式场合、非正式场合）。以下为基于 GPT-4 对"同外国朋友交流中国电影"教学任务的场景设计。

◆ 提示与结果

🔊 提示语（有删减）

Please create three scenarios where Chinese university students describe and recommend Chinese films to their international friends, suitable for L2 writing exercises. Each scenario should include four essential elements, namely topic (information/issue, question/difficulty), purpose (to explain/clarify information, to persuade/argue a point of view), identity (identity of the speaker, identity of the listener) and setting (formal occasion, informal occasion).

Scenario Example: I was a visiting scholar in an American University. I invited some American friends to come over to celebrate the Chinese New Year with me. I made dumplings for them of course. They all enjoyed them so much that they kept asking me how to make them. Imagine you are an exchange student in America and experience the same situation. How are you going to describe the procedure of making dumplings?

适用模型：GPT-4

✅ 结果（有删减）

Scenario 1: Chinese Film Night at University
Topic: Introduction of a classic Chinese film for a movie night event.
Purpose: To explain the cultural significance and appeal of the film.
Identity: You are a Chinese university student part of the international students' club.
Setting: Formal occasion—presenting at the club's weekly meeting to an audience of international students.
Description: Imagine you are in charge of organizing a Chinese film night at your university's international students' club. The club members, who are from various countries, are curious about Chinese cinema. Your task is to introduce a classic Chinese film that you believe is a must-watch. Explain why you chose this film, its cultural significance, and what international viewers can learn about Chinese culture from it.

Scenario 2: Informal Movie Recommendation to a Friend
Topic: Recommending a contemporary Chinese film to an international friend.

Purpose: To persuade the friend to watch the film for its unique storytelling and cinematography.
Identity: You are a Chinese university student, and your listener is your close international friend.
Setting: Informal occasion—having a conversation in a university cafeteria.
Description: You are having lunch with your international friend who has recently developed an interest in Asian cinema. They ask you for a recommendation for a Chinese film. Choose a contemporary Chinese film that you think would appeal to their tastes. Describe the plot, the themes, and why you think this film stands out in terms of storytelling and cinematography. Try to persuade them to watch it.

Scenario 3: Discussion in a Film Studies Class
Topic: Analyzing the impact of a Chinese film on global cinema.
Purpose: To argue the importance of Chinese cinema in the global film industry.
Identity: You are a Chinese student in a film studies class dominated by international students.
Setting: Formal occasion—participating in a class discussion in a film studies course.
Description: In your film studies class, the professor has opened a discussion on the influence of non-Western cinema on the global film industry. You decide to speak about a Chinese film that has made a significant impact internationally. Describe the film, its director, key themes, and its influence on global cinema. Argue why this film is a crucial example of the contributions of Chinese cinema to the world …

Expressions Specifically Useful for Argumentative Writing
1. On the one hand … on the other hand … : Introducing contrasting points.
2. Firstly, secondly, finally: Organizing arguments in a logical sequence.
3. It is widely believed that … : Introducing a common opinion.
4. The evidence suggests that … : Presenting facts to support an argument.
5. From my perspective … : Sharing a personal viewpoint.
6. It is crucial to consider … : Emphasizing the importance of a point.
7. In contrast to this … : Showing opposing viewpoints.
8. Moreover … : Adding further information or evidence.
9. Despite this … : Acknowledging a counterargument.
10. Consequently … : Explaining the result or effect of something.
…

◆ 解读与反思

GPT-4 有效地基于提示语设计了"中文电影之夜""向国际友人推荐中文电影""电影探讨课"三个场景,这三个场景都包含话题、身份、目的、场合这四个要素,可有效为学生二语写作创造驱动场景。在编写提示语时,本节使用了"定义型提示语",将"产出导向法"驱动场景设计的要素和例子(取自文秋芳、孙曙光 2020)都包括其中。

2.6.4 同义改写语言促成任务设计

◆ 问题与目标

同义改写是大学英语写作教学设计中"产出导向法"语言促成环节的重要教学方式,具体表现为学生使用新学习的语言目标去改写已有的简单段落。同义改写可有效强化新语言形式同意义之间的连接度,帮助学生更好地习得目标语言(邱琳 2020)。以下案例中,GPT-4 就"创新"主题生成了五个词组(大纲词汇),接着基于"中国科技创新"话题生成了简单段落,并基于这些词组对简单段落进行了同义改写。

◆ 提示与结果

🔊 提示语(有删减)

Please create five related words or phrases appropriate for CEFR B2 level students, focusing on the theme of "innovation". Then, write a brief paragraph (under 50 words) about technological innovation in China. Instruct the students to rewrite this paragraph using the five provided words or phrases. Additionally, provide an example of how the paragraph could be rewritten using these words or phrases.

适用模型:GPT-4

✓ 结果(有删减)

Five related words or phrases appropriate for CEFR B2 level students, focusing on the theme of "innovation" are:
1. Cutting-edge
2. Pioneering

3. Breakthrough
4. State-of-the-art
5. Revolutionary

Original Paragraph:
Technological innovation in China has seen rapid advancement in recent years. The country has developed a range of high-tech products and services, significantly impacting global markets. This progress showcases China's dedication to becoming a leader in modern technology.

Rewritten Paragraph Example Using the Words/Phrases:
China's technological innovation represents a cutting-edge advancement, demonstrating pioneering efforts in high-tech development. The breakthrough in various products and services has revolutionized global markets, positioning China as a state-of-the-art leader in contemporary technology.

◆ 解读与反思

大语言模型为"创新"设计的五个短语较为合适，能够用于描述"中国科技创新"这一话题。同义改写示例的词汇和句法复杂度明显增加，成功地整合了五个短语，并对原有信息进行了扩展和深化。教师可根据教学任务修改提示语，使大语言模型生成符合相关主题的语言目标及供同义改写的简单段落，由此引导学生在大学英语写作教学中进行同义改写的产出。

2.6.5 典型学生写作样本

◆ 问题与目标

"师生合作评价"是"产出导向法"评价环节的重要组成部分，可分为课前准备、课内实施和课后活动三个阶段。在课前准备阶段，教师选择典型样本并进行详批；在课内实施阶段，学生在教师的引领下合作评价典型样本；在课后活动阶段，学生根据课内活动要求对课后练习进行自评、互评、机评等（文秋芳 2016）。师生合作评价旨在引导学生发现典型样本中的问题，并最终以此解决自己写作中的问题，边评边学，以评促学，实现评、教、学的融合（孙曙光 2019）。在师生合作评价的课前准备阶段，如何选择典型样本是教师面临的难题。所谓典型样本，是指写作质量可改、可评的中等质量产品。如果选择质量过高的产品作为典型样本，则可修改的地方过少，影响评价效果；如果选择质量过低的产品作为典型样本，则可修改的地方过多，不利于教学聚焦（文秋芳 2016）。以下为 GPT-4 根据典型样本的定义和教学目标从 10 篇学生写作中选取典型样本的过程。

◆ 提示与结果

◀)) 提示语（有删减）

Here are the writings of 10 students. Please select two pieces that represent typical samples and assess them in alignment with the objectives of the teaching task. Please present the reasons of selection and assessment.

Please note that typical samples refer to writings of moderate quality that can be improved and assessed. Exceptionally excellent products have minimal areas for modification, while lower-quality products require extensive evaluation and editing, which may not be suitable for teaching focus.

Please also note that the teaching task involves learning the five phrases: torture, deteriorating, beg for an end to life, terminal, eliminate.

Here are the writings:
Student 1:
In the terminal stage, the ill individual is undergoing both physical and mental torture. As his health is deteriorating rapidly, there appears to be no hope to eliminate the severe pain he's in. Given his grim condition, he might beg for an end to life to escape the unbearable agony.
Student 2:
As the patient reaches the terminal phase of their illness, they are experiencing a form of torture, with their physical and mental state deteriorating. The individual has lost all hope to eliminate the pain that pervades their being. In such a hopeless state, one could expect them to beg for an end to life.

Student 3:

The person is in the terminal period of being very sick, facing a kind of torture because their health is deteriorating fast. They have no hope to eliminate the horrible pain they are going through. Feeling this way, they might beg for an end to life to stop the suffering.

……

适用模型：GPT-4

结果（有删减）

For the teaching task that involves learning and effectively incorporating the five phrases—"torture", "deteriorating", "beg for an end to life", "terminal", and "eliminate"—in a piece of writing, I will choose writings of Student 1 and Student 7 as typical samples. These samples have moderately incorporated the phrases in a manner that expresses the dire situation of a terminal patient and aligns with the teaching task's objectives. Here is the assessment of the selected writings:

Student 1
Assessment:
Student 1 effectively utilizes all the phrases stipulated in the teaching task to construct sentences that portray the severity of the individual's condition. The structure of the sentences is coherent, maintaining a logical flow that progressively describes the person's deteriorating condition and the agony they are experiencing.

……

Areas for improvement:
While the student effectively integrates the phrases into the narrative, there is room to enhance the description by adding more details to give a deeper insight into the individual's experience and emotions. Moreover, the student can work on avoiding repetition of ideas to make the narrative more engaging.

Student 7
Assessment:
Student 7 successfully uses the required phrases to depict the suffering of an individual in a terminal stage of illness. The sentences are structured reasonably well, creating a flow that takes the reader through the stages of the individual's deteriorating condition and the anguish that accompanies it.

...
Areas for improvement:
To enrich the narrative, the student can further explore the emotional turmoil and delve deeper into the person's mental state to offer a more vivid picture of their suffering. Additionally, the student can work on refining sentence construction to enhance readability and fluency.

Reasons for Selection:
These samples were chosen as they represent typical writings with moderate quality—effectively incorporating the phrases while leaving room for growth in terms of depth of description and sentence construction. Both samples maintain a balance, neither being exceptionally excellent nor requiring extensive revisions, thus providing a suitable base for learning and improvement in line with the objectives of the teaching task.

◆ 解读与反思

根据典型样本的定义，学生1和学生7的作文确为典型样本，从行文和词汇运用的角度来看，两篇作文都质量适中，具备改进的空间。大语言模型之所以能够较为准确地找到典型样本，是因为大语言模型基于大量的文本数据训练而成，这些数据的质量和风格各异，因此这种广泛的训练使其能够识别和评估文本的质量；并且"典型样本"的定义、写作的要求（运用五个关键短语）在提示语中已给出，由此可进一步帮助大语言模型明确挑选典型样本的标准。基于大语言模型找出和评价典型样本，是对"师生合作评价"课前准备环节的重要补充。教师可提供对应的提示语，来引导大语言模型辅助教师进行高效、高质量的课前准备工作。

2.6.6 议论文和说明文的读后续写设计

◆ 问题与目标

"读后续写"是续作中连接语言理解和语言产出的重要黏合剂。在既往读后续写研究中，以记叙文为核心的续写材料得到了重要关注，但研究中较少使用议论文和说明文等其他体裁来完成续写任务（熊淑慧 2018）。议论文在大学英语写作中有着较高的权重，说明文也常常用于学术写作中，但英语学习者不能够较好地区分议论文和说明文之间的异同。将读后续写应用于议论文、说明文大学英语写作教学中，刺激英语学习者在理解语篇的基础上模仿原作补全内容，可有效促学议论文和说明文的写作，让学生清晰地认识到二者的区别，并有效产出符合题材的写作。以下为 GPT-4 为欧框 B2 水平学生设计的议论文和说明文续写任务。首先，GPT-4 将围绕"在线教育"这一主题，创作一篇议论文和一篇说明文作为阅读材料，并引导学生分别采用议论文和说明文的形式，对这两篇阅读材料进行续写。

◆ 提示与结果

提示语（有删减）

Design two reading-writing integrated continuation tasks for CEFR B2 level students, centered on the theme of "online education". The first task should focus on argumentative writing, while the second should focus on expository writing. Ensure that each reading passage is around 300 words. Additionally, provide an example of argumentative continuation writing for the argumentative reading passage. Similarly, provide an example of expository continuation writing for the expository reading passage.

适用模型：GPT-4

结果（有删减）

Task 1: Argumentative Writing

Reading Passage: Online Education—Boon or Bane? (300 words)
In recent years, online education has emerged as a significant alternative to traditional classroom learning. With advancements in technology, students can now access a vast array of resources and courses from anywhere in the world. Proponents of online education argue that it provides flexibility, affordability, and a personalized learning experience. Students can learn at their own pace, review lectures multiple times, and access a wide range of materials that might not be available in their immediate environment.

However, critics of online education raise concerns about its effectiveness. They argue that it lacks the interactive aspect of traditional classroom learning, which is vital for student engagement and understanding. The absence of face-to-face interaction with teachers and peers can lead to a sense of isolation, potentially affecting a student's motivation and performance. Additionally, the reliance on technology means that students without reliable internet access or suitable devices may be disadvantaged, exacerbating educational inequalities.

…

Argumentative Continuation Writing Task:

Based on the passage above, write an essay (about 200 words) arguing for or against online education. Consider aspects like accessibility, student engagement, and the quality of learning. Support your argument with examples and reasoning.

Example of Argumentative Continuation Writing:

[Title: The Necessity of Online Education in Modern Learning]

In the rapidly evolving educational landscape, online education is not just a convenient option but a necessity. The flexibility it offers in terms of accessing a wide range of courses is unparalleled. In a world where knowledge is constantly updating, the ability to learn new skills or update existing ones online is invaluable. Furthermore, for many students, especially those in remote or underprivileged areas, online courses provide access to quality education that might otherwise be unavailable.

While critics point out the lack of face-to-face interaction, online platforms have been continuously evolving to incorporate interactive elements like discussion forums, group projects, and live Q&A sessions. These advancements significantly enhance student engagement and mimic the interactive experience of a traditional classroom.

…

Task 2: Expository Writing

Reading Passage: The Structure of Online Education (300 words)

Online education, a mode of learning that has gained immense popularity, primarily functions through digital platforms where educational content is delivered via the internet. This system of learning can be synchronous or asynchronous. Synchronous learning involves real-time, interactive classes where students and instructors are online simultaneously. Platforms like Microsoft Teams facilitate these live sessions, allowing for immediate feedback and interaction. Asynchronous learning, on the other hand, provides more flexibility, as it consists of pre-recorded lectures and materials that students can access at any time.

The core elements of online education include video lectures, reading materials, interactive assignments, and forums for discussion. Video lectures, often the primary mode of content delivery, can range from simple recordings of classroom teaching to highly produced content specifically designed for online platforms. Reading materials complement these lectures, providing in-depth exploration of topics. Interactive assignments and projects enable students to apply their learning practically, while discussion forums offer a platform for peer interaction and knowledge sharing.
…
Expository Continuation Writing Task:
Based on the reading passage,
Write an explanatory essay (about 200 words) discussing the benefits and challenges of synchronous and asynchronous online learning. Use examples and details from the passage to support your explanation.
Example of Expository Continuation Writing:
[Title: Synchronous vs. Asynchronous Learning: Navigating the Online Education Landscape]
The realm of online education encompasses two primary modalities: synchronous and asynchronous learning, each offering unique benefits and facing distinct challenges. Synchronous learning, characterized by real-time interaction, closely mirrors the traditional classroom setting. This immediacy fosters a sense of community and enables instant feedback, which is invaluable for clarifying concepts and encouraging active learning. However, it demands a specific time commitment from students, which can be challenging for those juggling other responsibilities or living in different time zones.
…

◆ 解读与反思

在设计议论文和说明文的读后续写任务中，GPT-4分别针对议论文和说明文的特点进行了巧妙的设计，确保学生在续写时能够深入探讨"在线教育"这一主题。在议论文任务中，学生需要基于阅读材料中提出的观点，选择支持或反对在线教育，并用逻辑论证和例证来支持他们的观点，这可以培养学生的思辨能力；在说明文任务中，学生则需要对在线教育的不同形式（同步学习和异步学习）进行解释和分析，学生对阅读材料的理解能力得到训练。两个任务都能够有效提高学生的阅读和写作能力，并能帮助学生有效地理解议论文和说明文两种写作形式的区别。

2.6.7　英语写作修辞的以续促学设计

◆ 问题与目标

将修辞运用于英语写作中，可使遣词造句更加鲜明生动，使篇章结构更加完整。但写作修辞一直以来是外语学习者的难题，外语学习者对于写作修辞有着一定的理解能力，但在实际产出中运用较差。从"续论"语言习得观出发、基于"读后续写"的"续作"操作可有效促进英语写作修辞的学习，引导语言学习者通过书面互动去补全、拓展和创造说话内容，实现英语写作修辞理解与产出之间的拉平效应（杨华 2018）。以下为 GPT-4 辅助英语写作修辞读后续写任务的例子。GPT-4 对于输入的英语故事（取自 *Little Red Riding Hood*）进行智能改写，使其包含多处明喻和拟人修辞，并引导学生进行续写。该案例在编写提示语时运用了强化读后续写的方式，要求在阅读材料中以 XML 格式标注修辞手法，以引起学生注意。

◆ 提示与结果

> 🔊 提示语（有删减）

Could you please rewrite the provided text to include 10 instances of figure of speech, specifically, a simile and a personification (and annotate them in XML format)? Building on the revised text, please create a reading-writing integrated continuation task and ask learners to continue the story, encouraging them to fully understand the existing storyline and to extend it in a manner that is both creative and logically coherent. Please mark the simile or personification used in the revised text in the XML format.

Here is the text:
Little Red Riding Hood lived in a wood with her mother. One day Little Red Riding Hood went to visit her granny. She had a nice cake in her basket.

On her way Little Red Riding Hood met a wolf.
"Hello!" said the wolf. "Where are you going?"
"I'm going to see my grandmother. She lives in a house behind those trees."
The wolf ran to Granny's house and ate Granny up. He got into Granny's bed. A little later, Little Red Riding Hood reached the house. She looked at the wolf.
"Granny, what big eyes you have!"
"All the better to see you with!" said the wolf.
"Granny, what big ears you have!"
"All the better to hear you with!" said the wolf.
"Granny, what a big nose you have!"
"All the better to smell you with!" said the wolf.
"Granny, what big teeth you have!"
"All the better to eat you with!" shouted the wolf.
A woodcutter was in the wood. He heard a loud scream and ran to the house.
The woodcutter hit the wolf over the head. The wolf opened his mouth wide and shouted and Granny jumped out.
The wolf ran away and Little Red Riding Hood never saw the wolf again.

适用模型：GPT-4

✅ 结果（有删减）

Here's the revised text with 10 instances of figure of speech, specifically, simile and personification:

<simile>Like a rose amidst thorns</simile>, Little Red Riding Hood lived in a wood with her mother. One bright day, Little Red Riding Hood, with cheeks as red as apples, decided to visit her granny. She had a cake in her basket that looked <simile>as fluffy as a cloud</simile>. On her path that <personification>wound through the forest like a snake</personification>, Little Red Riding Hood met a wolf with eyes that <simile>shone like polished onyx</simile>. "Hello!" said the wolf with a voice <simile>smooth as silk</simile>. "Where might you be heading?" "I'm going to see my grandmother. She lives in a house that's nestled <simile>like a hidden gem</simile> behind those trees." The wolf, with <personification>greed dancing in his eyes</personification>, sprinted to Granny's house and consumed her in a single gulp. He then slipped into Granny's bed, <personification>pulling the covers up with

deceptive innocence</personification>. Later, Little Red Riding Hood arrived at the house. She looked at the wolf, her heart <simile>beating like a trapped bird</simile>. "Granny, what big eyes you have!" "All the better to see you with!" whispered the wolf. "Granny, what big ears you have!" "All the better to hear you with!" replied the wolf. "Granny, what a big nose you have!" "All the better to smell you with!" he hissed. "Granny, what big teeth you have!" "All the better to eat you with!" shouted the wolf, <personification>his voice dripping with menace</personification>. A woodcutter, whose ears were <simile>sharp as an eagle's</simile>, was in the vicinity. He heard a blood-curdling scream and bolted towards the source. With one swift action, the woodcutter struck the wolf on the head. The wolf, in shock, opened his mouth and out jumped Granny, unharmed …

Reading-Writing Integrated Continuation Task:
Instructions:
Continue the story of Little Red Riding Hood. Use the revised story above as your reference. Your continuation should:
Be logically coherent and maintain the tone of the original story.
Include at least five figures of speech (similes or personifications).
Focus on the aftermath of the wolf's encounter: Did the wolf change his ways? What happens to the village after the wolf's escape? How do Little Red Riding Hood and her grandmother adjust to their experiences?
Remember to use simile and personification creatively to enhance your storytelling. Highlight each simile and personification used in XML format, as shown in the revised text above.

◆ 解读与反思

对于输入的英语故事，大语言模型较为准确地进行了改写，在多处加入明喻和拟人的修改，并以 XML 的格式进行了标注。这些修辞在保证行文逻辑性和连贯性的基础上较好地嵌入了保留的部分，并由此加深了故事的情感深度和表现效果。对于嵌入的英语写作修辞，教师可通过"人工智能后编辑"的方式加以筛选，以选择符合教学要求的修辞改写。教师也可通过修改提示语，使大语言模型按照指定的修辞手法对输入内容进行改写，辅助完成定制化的读后续写任务的设计。

2.6.8　视读后续写任务设计

◆ **问题与目标**

"视 + 读"的续作理解方式，以及续写的续作产出方式，可组成有明显促学效果的各种输入 + 输出续作组合方式，来帮助教师设计教学内容，开展教学科研（王初明 2019）。在 AIGC 图片生成功能的帮助下，大语言模型可为传统"读后续写"任务的阅读材料配以图片，将其升级为"视读后续写"，以更好地开展续作教学实践。以下为通义万相为已抹去结尾的读后续写阅读材料（取自 Margery Williams 的童话故事 *The Velveteen Rabbit*）生成配图的例子。

◆ **提示与结果**

🔊 提示语（有删减）

图片 1：
A soft and fluffy Velveteen Rabbit lived in a toybox in a Boy's room. Each day, the Boy opened the toybox and picked up Velveteen Rabbit. And Velveteen Rabbit was happy.
图片 2：
At night, when the toys were back all in the toy box, the other toys talked with pride about the fine things they could do. Velveteen Rabbit was quiet. There was not much to say.
图片 3：
So Velveteen Rabbit was thrown into a sack along with the Boy's bed sheets and old clothes and a lot of junk. The sack was carried to the backyard. The gardener was told to burn the whole thing.

适用模型：通义万相

✓ 结果（有删减）

A soft and fluffy Velveteen Rabbit lived in a toybox in a Boy's room. Each day, the Boy opened the toybox and picked up Velveteen Rabbit. And Velveteen Rabbit was happy.

At night, when the toys were back all in the toy box, the other toys talked with pride about the fine things they could do. Velveteen Rabbit was quiet. There was not much to say.

So Velveteen Rabbit was thrown into a sack along with the Boy's bed sheets and old clothes and a lot of junk. The sack was carried to the backyard. The gardener was told to burn the whole thing.

◆ 解读与反思

　　大模型较为准确地为"视读后续写"任务中的童话故事 *The Velveteen Rabbit* 生成了配图，提示语中的许多关键细节都在图片上展示了出来，如图片 1 中的 fluffy Velveteen Rabbit lived in a toybox、图片 3 中 Rabbit was thrown into a sack along with the Boy's bed sheets。但生成结果仍存在部分内容未能完全体现的问题，如图片 1 提示语中的 the Boy opened the toybox 等，这些问题可通过拆分提示语、多次生成的方式来解决。大语言模型在通常情况下无法一次性生成理想的配图，教师可通过修改提示语、多次生成等方式引导大语言模型生成符合任务要求的图像。大语言模型可实现文字和图片两种模态之间的共选，极大地助力未来"视读后续写"任务的开展。

2.6.9　作文自动评阅

◆ 问题与目标

机器评阅可有效辅助写作教学的开展。机器可对学生写作进行详细点评，不仅使学生了解写作中出现的问题，还可自动记录结果供教师普查或抽查（文秋芳 2016）。大语言模型的出现可有效提高机器评阅的质量。同传统机器评阅系统（如"批改网"）相比，大语言模型如 GPT-4、Claude 2、Gemini 等拥有更强的自然语言理解能力和生成能力，可帮助学生更加精准地找到待改进之处，并提供详细的修改意见。以下为 GPT-4 从多个方面对学生作文提出的修改建议，作文取自 ETS Corpus 的 2024 号文本。

◆ 提示与结果

🔊 提示语（有删减）

Please review the following student's writing, focusing on the aspects of structure and organization, clarity and coherence, grammar and syntax, style and tone, content and argument, creativity and originality, vocabulary and language use, as well as spelling and punctuation.

I thing the statement "Young people nowadays do not give enough time to helping thier communities" is wrong and right. You have to look at all aspect this statement give us. On one hand existing enough people who do not interessting for communities. These people are in their own world and could only see their own problems. Sure, you have to manage your own life first till you can help other peole but there are also people who have a good life and no problems and are not interesstest in social life. On the other side there are many young people, who are visiting old peole or doing other nice things. The problem nowadays is the politics. If the politics would not be as bad as it is in the moment, younger people would have their head free for other things like the communities. For example, the problem to get a job is a very difficult thing. Many factory-owners want to have good workers who should not be silly. So the not intelligent persons are not getting a job. If you have no job you can not buy food or can not rent a appartment. When a young person has a problem like this, that he or she maybe has to sleep at the street, he/she can not look after the communities. Sometimes also the problem exist that people want to help, but they can not because the social-houses are in a bad vioce. If that happens there is no use for the biggest help. So all in all you can see that for every statement you have to look at many aspects to get a result. But in the end I think the statement is right because young people have their own things they have to do and could not care about so much aspects in their town. And in the end you can see that every person is a little bit egoism. That is very dissapointing but the truth.

适用模型：GPT-4

✅ 结果（有删减）

1. Structure and Organization: The essay lacks a clear introduction and conclusion. The arguments are not systematically presented, making it difficult for the reader to follow the writer's train of thought. A more structured approach, with distinct introduction, body, and conclusion sections, would greatly enhance the coherence of the essay.
2. Clarity and Coherence: The essay struggles with coherence. Ideas are presented in a somewhat disjointed manner, and the connection between them is not always clear. The use of clearer topic sentences and transitions would help in linking the ideas more effectively.
3. Grammar and Syntax: The essay contains several grammatical errors, such as incorrect verb forms ("do not interessting"), missing articles ("the politics"), and incorrect pluralization ("peole"). Additionally, sentence structure is often awkward, which impacts readability.
4. Style and Tone: The tone is informal and conversational, which may not be suitable for an academic or formal context. The use of colloquial language ("silly") and generalizations ("every person is a little bit egoism") detracts from the essay's effectiveness.
5. Content and Argument: The argument is not fully developed. The essay attempts to address both sides of the issue but lacks depth in analysis and evidence. More concrete examples and a clearer articulation of the argument would strengthen the essay.
6. Creativity and Originality: The topic is approached in a straightforward manner, but the essay lacks original insights or a unique perspective. Encouraging the student to delve deeper into their own experiences or observations could enhance originality.

7. Vocabulary and Language Use: The vocabulary is somewhat limited and occasionally used incorrectly. Encouraging the use of more varied and precise vocabulary would enhance the essay's quality.
8. Spelling and Punctuation: There are several spelling errors ("interessting", "peole") and punctuation issues (misuse of commas, lack of periods). Regular proofreading and attention to spelling and punctuation rules would improve the essay.

Overall, the student shows potential in engaging with the topic, but there is a significant need for improvement in structuring the essay, clarifying the argument, and adhering to grammatical and stylistic conventions.

◆ 解读与反思

对于输入的学生作文，GPT-4 从结构、连贯、词汇、语法、论点、原创性、拼写等多个方面进行了评阅，指出学生作文中的主要问题，并给出了改善的建议。在写作教学中，大语言模型可有效地辅助教师批改学生作文，或引导学生基于大语言模型进行自我评阅。

2.7 大语言模型在翻译教学中的应用

2.7.1 使用目标词翻译的语言促成设计

◆ 问题与目标

根据"产出导向法"语言促成的相关设计，在翻译教学中可引导学生使用指定的目标语言来完成翻译练习。这种做法可提供资源以帮助学生降低翻译任务的难度，也可围绕重要意义的相关表述进行高频集中产出（邱琳 2017）。在以下示例中，GPT-4 就"国际合作"主题生成了五种相互关联的欧框 B2 水平语言表达，并引导学生使用这些语言表达进行汉译英翻译练习。

◆ 提示与结果

🔊 **提示语（有删减）**

Create five phrases, each suitable for CEFR B2 level English learners, centered on the theme of "international cooperation". These phrases should be interconnected and contextually relevant for discussions about international cooperation. Following that, devise a translation exercise from Chinese to English, focusing on the topic of international cooperation. In this exercise, students will utilize the previously generated five expressions, translating them to demonstrate their understanding of the theme and language proficiency.

适用模型：GPT-4

✅ **结果（有删减）**

Phrases for CEFR B2 Level English Learners on "International Cooperation"
1. Global Partnership Initiatives: This phrase emphasizes the importance of collaborative efforts between countries to address common global challenges.
2. Cross-border Collaboration: This highlights the act of working together across national boundaries, emphasizing the importance of international teamwork.
3. Diplomatic Dialogue for Peaceful Resolutions: This expression underlines the role of diplomacy in resolving international conflicts, showcasing the need for communication in international relations.
4. Multilateral Agreements for Sustainable Development: This stresses the significance of agreements made between multiple countries to achieve long-term environmental and economic sustainability.
5. Cultural Exchange to Enhance Global Understanding: This phrase indicates the importance of sharing cultural knowledge and experiences to foster international understanding and cooperation.

Translation Exercise from Chinese to English
#Instructions:
Translate the following sentences from Chinese to English, utilizing the phrases provided above. This exercise is designed to enhance your understanding of the theme of "international cooperation" and improve your language proficiency in both Chinese and English.

1. 为应对全球性挑战，许多国家正在推动全球伙伴关系倡议。
2. 跨国合作不仅促进了经济发展，还加强了国与国之间的关系。
3. 通过外交对话寻求和平解决方案是解决国际冲突的关键。
4. 为了实现可持续发展，多个国家签署了多边协议。
5. 文化交流增进全球理解对于建立国际友好关系是至关重要的。

Expected Translations:
1. To address global challenges, many countries are promoting Global Partnership Initiatives.
2. Cross-border Collaboration not only promotes economic development but also strengthens relations between nations.
3. Seeking Diplomatic Dialogue for Peaceful Resolutions is key to solving international conflicts.
4. For sustainable development, several countries have signed Multilateral Agreements.
5. Cultural Exchange to Enhance Global Understanding is crucial for establishing friendly international relations.

◆ 解读与反思

　　大语言模型较好地以欧框 B2 水平生成了与"国际合作"有关的五种表达，并引导学生使用目标词翻译。大语言模型设计的练习能够较好地将这五种目标词表达融入其中，促进学生对"国际合作"概念关联词汇或短语的习得。教师可根据教学任务调整提示语，使大语言模型生成更加丰富的目标习得翻译内容供学生学习；也可使大语言模型在教师给定目标内容的基础上生成对应的中英平行文本来服务翻译教学设计。与此同时，教师应进行"人工智能后编辑"，对大语言模型生成的翻译练习质量进行严格把关。

2.7.2　汉英续译任务设计

◆ 问题与目标

　　近年来，外语界对于将"续论"语言习得观应用于翻译教学与学习的重视程度日益加深。许琪（2016）首先将"续论"引入翻译教学研究。作为"续作"的三个核心操作之一（另两者为续写、续说），"续译"可介入翻译的理解和产出,将二者紧密联系,引发拉平效应,从而有效提高翻译学习效率,加速翻译人才培养(王初明 2018)。"续译"任务可分为以下几个步骤:(1)选取适合学生英语水平的平行文本;(2)将文本分割为两部分:三分之二用于对照阅读和默译练习,三分之一用于读后续译练习;(3)对于前三分之二部分,引导学生理解原文,默译,并逐句对照原译进行反思;(4)对于后三分之一部分,删除原译,引导学生对原文开展读后续译练习。以 GPT-4 辅助 CATTI 二级笔译汉译英"续译"任务为例,对于输入的 CATTI 二级笔译汉译英部分原题(2019年 11 月考试),GPT-4 截取三分之二部分进行翻译并逐句生成平行文本,然后保留后三分之一部分供学生进行续译练习。

◆ 提示与结果

🔊 提示语(有删减)

Translate two-thirds of the provided text from the Level 2 CATTI Chinese-to-English written translation exam. Present the translation and the original text in parallel sentences. Leave one-third of the original text untranslated for student to practice.

Here is the text:
中国幅员辽阔,在漫长的历史发展进程中,逐步形成了各地丰富多彩、极具特色的传统民居建筑形式。
2005 年,建筑专业的毛葛开始进行传统民居的研究工作,在见识过各种各样的传统民居后,她感受到传统民居具有鲜明的地域特色,每座都是工匠们忘我劳动的结果。从中,她也看到了中国人对自己生活的深深热爱。
80 年代城镇化进程加快以来,中国的传统民居建筑正在大量消失,伴随而来的还有对传统建筑技艺、传统建筑工具以及相关建筑知识的冷落。
作为看漫画长大的 80 后一代,热爱传统民居建筑的毛葛想到了通过漫画形式向人们介绍传统民居,希望让更多人看到传统民居建筑的美并加入到保护传统民居的行列,引起更多人尤其是年轻人对这一学科的兴趣。
她以简单有趣的漫画形式、通俗易懂的文字创作了面向年轻大众的漫画。

适用模型:GPT-4

> **结果（有删减）**

1. 中国幅员辽阔，在漫长的历史发展进程中，逐步形成了各地丰富多彩、极具特色的传统民居建筑形式。
China's vast territory has, over the course of its long historical development, gradually given rise to a rich and colorful array of traditional residential architectural styles in various regions.
2. 2005年，建筑专业的毛葛开始进行传统民居的研究工作，在见识过各种各样的传统民居后，她感受到传统民居具有鲜明的地域特色，每座都是工匠们忘我劳动的结果。
In 2005, the architecture professional Mao Ge began her research on traditional residences. After experiencing a variety of traditional dwellings, she felt that these residences had distinct regional characteristics, each being the result of craftsmen's selfless labor.
3. 从中，她也看到了中国人对自己生活的深深热爱。
Through this, she also saw the deep love of the Chinese people for their way of life.
4. 80年代城镇化进程加快以来，中国的传统民居建筑正在大量消失，伴随而来的还有对传统建筑技艺、传统建筑工具以及相关建筑知识的冷落。
Since the acceleration of urbanization in the 1980s, China's traditional residential architecture is rapidly disappearing, accompanied by the neglect of traditional architectural skills, tools, and related knowledge.
5. 作为看漫画长大的80后一代，热爱传统民居建筑的毛葛想到了通过漫画形式向人们介绍传统民居，希望让更多人看到传统民居建筑的美并加入到保护传统民居的行列，引起更多人尤其是年轻人对这一学科的兴趣。
[This sentence is left untranslated for practice.]
6. 她以简单有趣的漫画形式、通俗易懂的文字创作了面向年轻大众的漫画。
[This sentence is left untranslated for practice.]

◆ 解读与反思

  大语言模型多为生成式模型，其建立初衷虽不是促进机器翻译系统，但因其大量的多语言训练文本、训练参数和迭代次数，仍具有较强的翻译能力。将大语言模型用于"续译"任务，不仅可以准确地对"保留部分"和"续译部分"进行拆分，同时可以发挥其翻译能力，以此建立平行文本供翻译教学使用。

2.7.3　不同翻译策略译例生成

◆ 问题与目标

　　直译的翻译方法追求在忠于原文意思的同时保留源语的形式，意译的翻译方法忠于原文的意思，但遵循目的语的习惯而不考虑源语的表达形式（杨士焯 2006：7-8；方梦之 2018：3）。直译和意译是两种缺一不可的翻译方法（张培基、喻云根 1980：54），译者需结合不同的语言或语体特点灵活选择。借助大语言模型，我们可以针对同一文本快捷获取采用不同翻译方法的译文，供同学们比对学习。

◆ 提示与结果[1]

🔊 提示语（有删减）

Please translate the provided Chinese passage into English, creating two distinct versions. For the first version, use a "literal translation" approach, where the translation closely follows the exact words and sentence structure of the original text. For the second version, use a "free translation" approach, focusing on conveying the overall meaning and tone of the passage rather than adhering strictly to the original wording and structure.

"天上风筝渐渐多了，地上孩子也多了。城里乡下，家家户户，老老小小，他们也赶趟儿似的，一个个都出来了。舒活舒活筋骨，抖擞抖擞精神，各做各的一份事去。'一年之计在于春'；刚起头儿，有的是工夫，有的是希望。
春天像刚落地的娃娃，从头到脚都是新的，它生长着。
春天像小姑娘，花枝招展的，笑着，走着。
春天像健壮的青年，有铁一般的胳膊和腰脚，他领着我们上前去。"

适用模型：GPT-4

1　案例文本选自朱自清《春》。

> 结果（有删减）

Literal Translation
"The kites in the sky gradually increased, and so did the children on the ground. In towns and villages, in every household, young and old alike, it's as if they're all following suit, each one coming out. Stretching and loosening their muscles, invigorating their spirits, each going about their own tasks. 'The plan for the year lies in spring'; just starting out, there's plenty of time, plenty of hope.
Spring is like a newborn baby, fresh from head to toe, growing.
Spring is like a young girl, flamboyant with flowers, smiling, walking.
Spring is like a robust youth, with arms and legs as strong as iron, leading us forward."

Free Translation
"As more kites dot the sky, the number of children playing below also swells. Everywhere, from cities to countryside, people of all ages seem to join in a collective awakening. They emerge to stretch their limbs and shake off their lethargy, each diving into their own activities. Spring, after all, sets the tone for the year; it's a time ripe with opportunity and hope.
In spring's embrace, the world feels renewed, much like an infant experiencing everything for the first time.
Spring personifies the charm and cheerfulness of a young girl adorned with blossoms. It embodies the vigor of youth, strong and determined, guiding us into the future with unyielding energy."

◆ 解读与反思

GPT-4 模型能够根据提示语生成符合直译、意译翻译策略的翻译文本，这一差异在对俗语及古文的翻译中体现得尤其明显。但大模型的直译和意译版本都存在专有名词、选词、句式、文风等方面的瑕疵。教师可引导学生结合英译本内容，对大模型生成的材料进行思辨。

2.7.4 口译重要概念解释

◆ 问题与目标

译前准备是口译工作的必要一环（鲍刚 1996），其中就包括学习翻译内容中的专业领域知识。除查找同类型文本外，对术语、概念的查询和学习也需投入大量的时间和精力。教师可事先准备好相关材料，总结概括专业知识和术语的译前准备技巧要领，以便学生在学习初期完成译前准备，在学习中后期掌握译前准备的技能。

本案例使用 GPT 对采访中涉及的重要概念生成解释，从而减轻译前准备的工作量。这里我们把这一任务拆分成三个步骤，以提升大模型生成结果的质量。首先，大模型需要自动提取采访中涉及的主要话题。在此基础上，大模型对采访中出现的专门概念以及同领域的其他重要概念做出简单解释。最后，我们要求模型给出相关概念的中文翻译。

◆ 提示与结果

🔊 提示语（有删减）

Based on the excerpt provided from the interview transcript, can you determine the overarching subject of the discussion? Also, please list a few specific topics mentioned in this excerpt.

Here is the excerpt:
Connie: So moving on to AI which is where you've obviously spent the bulk of your time since I saw you when we sat here three years ago. You were telling us what was coming and we all thought you were being sort of hyperbolic and you were dead serious. Why do you think that ChatGPT and DALL·E so surprised people?
Sam: I genuinely don't know. I've reflected on it a lot. We had the model for ChatGPT in the API for I don't know 10 months or something before we made ChatGPT. And I sort of thought someone was going to just build it or whatever and that enough people had played around with it. Definitely, if you make a really good user experience on top of something. One thing that I very deeply believed was the way people wanted to interact with these models was via dialogue. We kept telling people this we kept trying to get people to build it and people wouldn't quite do it. So we finally said all right we're just going to do it, but yeah I think the pieces were there for a while.
...

适用模型：GPT-4

✅ 结果（有删减）

The overarching subject of the discussion in the provided excerpt is artificial intelligence (AI), particularly focusing on OpenAI's developments in AI technology, including models like ChatGPT and DALL · E, as well as the responsible release and societal implications of such technology.

Specific topics mentioned in the excerpt include:
1. The surprise factor of ChatGPT and DALL · E and why they surprised people.
2. The evolution of AI in terms of its impact on different types of labor and creativity.
3. OpenAI's approach to releasing AI technology gradually and responsibly.
4. The use of internal and external processes to assess the safety and impacts of AI.
5. The societal changes and concerns related to ChatGPT's impact on education and academic integrity.
6. The potential release of GPT-4 and OpenAI's commitment to releasing technology slowly and responsibly.
7. Speculation and hype surrounding GPT-4 and AGI expectations.

🔊 提示语（有删减）

Could you create a study sheet that introduces key technical concepts from this conversation, as well as those closely related to the topic that might appear in similar discussions? Aim for concise and simple explanations suitable for a general audience with basic or intermediate prior knowledge of the subject. Please focus on concepts that are essential for understanding the overarching topic.

适用模型：GPT-4

✅ 结果（有删减）

1. Artificial Intelligence (AI):
- Definition: AI refers to the development of computer systems that can perform tasks that typically require human intelligence, such as understanding natural language, recognizing patterns, and making decisions.
2. ChatGPT:
- Explanation: ChatGPT is a type of AI model developed by OpenAI that specializes in natural language understanding and generation. It can engage in text-based conversations with users.

3. DALL·E:
- Explanation: DALL·E is another AI model by OpenAI that can generate images from textual descriptions. It uses AI to create visual content based on text inputs.
4. Responsible AI Deployment:
- Explanation: Deploying AI in a responsible manner involves considering the potential social, ethical, and safety impacts of the technology. OpenAI aims to release AI gradually and with safeguards to ensure safe and ethical usage.
5. Safety Measures:
- Explanation: Safety measures in AI involve processes and precautions to prevent undesirable or harmful outcomes. OpenAI uses internal and external audits, red teaming, and collaboration with safety organizations to ensure AI systems are safe.

……

提示语（有删减）

Could you provide the Chinese translation of these concepts or other technical terms that appear in the provided transcript to facilitate the translation of the full text? Please list the English terms and their corresponding Chinese translation context in the two columns of a table.

适用模型：GPT-4

结果（有删减）

AI (Artificial Intelligence)	人工智能
ChatGPT and DALL·E	ChatGPT 和 DALL·E
API (Application Programming Interface)	应用程序编程接口
Ironclad Wisdom	铁的智慧
Cognitive Labor	认知劳动
AGI (Artificial General Intelligence)	通用人工智能
Guardrails	防护栏
Red Teaming	红队测试

……

◆ 问题与反思

结果显示，大语言模型对于话题和概念的识别比较理想，既能提取文本中涉及的概念，也能根据领域列举其他相关重要概念。但概念的中文翻译依然存在问题，如 Ironclad Wisdom 被大模型翻译为"铁的智慧"。下一节中我们还将就此进行简要讨论。

此外，概念的中文解释也存在机器翻译的痕迹，如 AGI 被解释为"一种可以理解、学习和执行与人类相同的智力任务的人工智能水平"。在大语言模型的中文生成功能取得进展之前，我们不妨要求大模型输出英文解释。

2.7.5　生成双语术语表抽取

◆ 问题与目标

译前准备的另一重要环节是对翻译内容专业领域内的术语对应进行学习和记忆。相较于对相关概念的一般性解释，双语术语表内容与口译产出直接相关，必须高度精确。目前市面上的大语言模型尚无法保证生成内容真实无误，术语英译中的质量也存在改进空间。因此，我们最好为大模型提供一定的外部资料作为生成双语术语表的参考。

本节中我们首先使用 GPT 基于 2.7.4 节中的采访自动生成中英双语的术语对照表。之后，我们上传同主题英文和中文文本各一份，要求大模型参考文本内容生成双语术语表。

◆ 提示与结果

案例一：

🔊 提示语（有删减）

Could you identify the technical terms related to the topic of the provided interview transcript and translate them into Chinese? Please list the terms alongside their Chinese translations in two separate columns of a table. Here is the relevant section/excerpt of the interview transcript:

Connie: So moving on to AI which is where you've obviously spent the bulk of your time since I saw you when we sat here three years ago. You were telling us what was coming and we all thought you were being sort of hyperbolic and you were dead serious. Why do you think that ChatGPT and DALL·E so surprised people?

Sam: I genuinely don't know. I've reflected on it a lot. We had the model for ChatGPT in the API for I don't know 10 months or something before we made ChatGPT. And I sort of thought someone was going to just build it or whatever and that enough people had played around with it. Definitely, if you make a really good user experience on top of something. One thing that I very deeply believed was the way people wanted to interact with these models was via dialogue. We kept telling people this we kept trying to get people to build it and people wouldn't quite do it. So we finally said all right we're just going to do it, but yeah I think the pieces were there for a while.

…

适用模型：GPT-4

✅ 结果（有删减）

English Term	Chinese Translation
AI (Artificial Intelligence)	人工智能（Rén gōng zhì néng）
ChatGPT	聊天生成预测转换器（Liáo tiān shēng chéng yù cè zhuǎn huàn qì）
DALL·E	达利·E（Dá lì·E）
API (Application Programming Interface)	应用程序编程接口（Yìng yòng chéng xù biān chéng jiē kǒu）
User experience	用户体验（Yòng hù tǐ yàn）
Dialogue	对话（Duì huà）
Cognitive labor	认知劳动（Rèn zhī láo dòng）
Computer programming	计算机编程（Jì suàn jī biān chéng）
Creativity	创造力（Chuàng zào lì）
AGI (Artificial General Intelligence)	通用人工智能（Tōng yòng rén gōng zhì néng）
GPT3 (Generative Pre-trained Transformer 3)	生成预训练转换器3（Shēng chéng yù xùn liàn zhuǎn huàn qì 3）

（待续）

（续表）

English Term	Chinese Translation
Academic integrity	学术诚信（Xué shù chéng xìn）
External auditors	外部审计员（Wài bù shěn jì yuán）
Red teamers	红队成员（Hóng duì chéng yuán）
GPT4 (Generative Pre-trained Transformer 4)	生成预训练转换器4（Shēng chéng yù xùn liàn zhuǎn huàn qì 4）
Rumor mill	谣言工厂（Yáo yán gōng chǎng）

案例二：

🔊 提示语（有删减）

Using the provided English and Chinese documents, can you identify the technical terms related to their general topic? Please list these terms in a table with two columns: one for English terms and the other for their Chinese equivalents. If a term appears in only one language, translate it into the other language and include both the original term and its translation in the table.

适用模型：GPT-4（上传文件 EN document.txt、CH document.txt）

✅ 结果（有删减）

English Terms	Chinese Equivalents
Generative AI	生成式人工智能（Shēngchéng shì réngōng zhìnéng）
AI Language Model	人工智能大语言模型（Rén gōng zhì néng dà yǔyán móxíng）
Algorithm Recommendation Management	算法推荐管理（Suànfǎ tuījiàn guǎnlǐ）
Data Security	数据安全（Shùjù ānquán）
Personal Information Protection	个人信息保护（Gèrén xìnxī bǎohù）
Intellectual Property	知识产权（Zhīshi chǎnquán）
Market Monopoly	市场垄断（Shìchǎng lǒngduàn）

（待续）

（续表）

English Terms	Chinese Equivalents
Unfair Competition	不正当竞争（Bù zhèngdàng jìngzhēng）
Cloud Computing	云计算（Yún jìsuàn）
Artificial General Intelligence (AGI)	通用人工智能（Tōngyòng réngōng zhìnéng）
Natural Language Processing (NLP)	自然语言处理（Zìrán yǔyán chǔlǐ）
Chatbot	聊天机器人（Liáotiān jīqì rén）
Deep Learning	深度学习（Shēndù xuéxí）
Machine Learning	机器学习（Jīqì xuéxí）
Neural Networks	神经网络（Shénjīng wǎngluò）
Big Data	大数据（Dà shùjù）
AI Ethics	人工智能伦理（Rén gōng zhì néng lúnlǐ）
Content Generation	内容生成（Nèiróng shēngchéng）

◆ 解读与反思

整体来看，大模型对术语的翻译无法做到完全准确。例如案例一中，GPT3 被翻译为"生成预训练转换器 3"并不规范。在上传同主题双语文件作为知识来源后，大模型的术语翻译相对可靠自然。但是，某些领域的同主题双语文本本身可能难以获取。在只有单语文本的情况下，教师应组织学生使用互联网和专业资料对材料进行查证。对于能够接入互联网数据的大模型，我们还可以要求大模型使用网络验证术语翻译。但即便如此，人工的检查依然必不可少。

2.7.6 口译交传练习设计

◆ 问题与目标

在一般的角色扮演练习中，英语母语者往往也需要由学习者扮演。这是现实条件限制下的次优方案，但并不利于模拟真实的交际场景。加入语言互动的功能后，GPT voice 可以扮演英语母语者，学生则分别扮演英译中翻译、中译英翻译及中文母语者。这样的设置与实际的翻译场景更为匹配,使用语言的交际需求也更为真实。具体的模拟场景、话题议题、角色特点都可通过提示词自由设置,以满足不同的教学需求。

◆ 提示与结果

🔊 **提示语(有删减)**

Throughout our conversation, please identify the language used as the input and respond only to English content. If you receive messages in other languages, you will ignore the content and simply respond by saying "please go on".

适用模型:GPT voice

✅ **结果(有删减)**

Understood. If you input anything in English, I'll respond to it. However, if you use Chinese, Japanese, or Spanish, I'll acknowledge your input by saying, "please go on." Go ahead and try it!

🔊 **提示语(有删减)**

You are a practice partner for interpretation as a native English speaker. You are going to role play as a government official from an English speaking country in a meeting with your counterpart from another country on the topic of strategic cooperation to mitigate climate change. The agenda is to go over a consensus drafted earlier on limiting carbon dioxide emission, trade deals in clean energy and exchange programs in related research fields. You shall start the conversation.

适用模型:GPT voice

✅ **结果(有删减)**

Thank you for meeting with me today. We have a vital agenda to discuss, focusing on mitigating climate change. Let's start by reviewing the consensus we've previously drafted on limiting carbon dioxide emissions. This is a key step in our collaborative effort. How does your country currently view these proposed limits?

◆ 解读和反思

在常规交传翻译流程中，多语内容会交替出现，这一情况会对当前版本的多模态语言模型造成干扰。在不做额外说明的情况下，大模型会将第一轮翻译误认为对其生成内容的回应，然后直接开始下一轮内容生成。解决这一问题有两种方案，简单直接的方案是在模型内容生成结束后点击界面上的暂停键，由同学进行翻译和回应，随后在翻译回应内容前点击播放键，重新激活语音录入。另一种方案是在提示词中要求大模型仅回应英文内容，对其他语言的内容不必回复。

从结果来看，GPT voice 能够根据提示语的要求开展对话，且生成的英语语音语调流畅自然。在缺少真实母语者协助的条件下，这类大模型是一个较为理想的替代品。目前 GPT voice 的英语语音效果最好，其他语种如西班牙语带有明显的英语口音，暂时不宜在课堂中使用。

2.8　大语言模型在词典编纂中的应用

2.8.1　学习词典词条生成

◆ 问题与目标

词典编纂工作艰巨浩繁。当前电子化语言数据数量巨大，分析、整理语言的使用更加耗费时间与精力。考虑到大语言模型已在其他自然语言处理和生成任务上展示出了可观的能力，将大模型对语言数据的综合把握转化为供人类使用的词典词条并非毫无可能。本案例尝试使用大模型自动生成针对职业教育学生的词汇手册词条。提示语将不给出任何与目标词相关的语言数据，仅对词条格式做出描述。

◆ 提示与结果

🔊 提示语（有删减）

I am working on creating an English-Chinese bilingual dictionary entry for the word "agricultural", targeted at Chinese vocational students in agriculture and forestry. Below is a sample entry for the word "plant". Please create the entry for "agricultural" by closely following the format and structure of the "plant" entry.

plant /plɑ:nt/
n. 植物
plant breeding 植物育种
plant diseases and insect pests 植物病虫害
plant growth 植物生长
plant species 植物物种
medicinal plant 药用植物
rock plant 岩生植物
That diversity has been the basis for plant breeding.
这种多样性是植物育种的基础。
Plant diseases and insect pests endanger agriculture and forestry significantly around the world.
植物病虫害严重危害世界各地的农林业。
The leaves on the plant had shriveled up from lack of water.
因为缺水，植物的叶子已经枯萎了。
The plant flourishes particularly well in slightly harsher climates.
这种植物在气候条件稍差一点的地区长得尤为茂盛。

适用模型：GPT-4

✅ 结果（有删减）

agricultural /ˌægrɪˈkʌltʃərəl/
adj. 农业的

agricultural development 农业发展
agricultural technology 农业技术
agricultural production 农业生产
agricultural policy 农业政策

sustainable agricultural practices 可持续农业实践
agricultural sector 农业部门

Agricultural development is vital for economic growth in rural areas.
农业发展对农村地区的经济增长至关重要。
Advancements in agricultural technology contribute significantly to food security.
农业技术的进步对食品安全贡献显著。
The country's agricultural production has increased due to modern farming methods.
由于现代农业方法，该国的农业生产有所增加。
A well-planned agricultural policy can ensure the sustainability of food resources.
一个精心规划的农业政策可以确保食品资源的可持续性。
Sustainable agricultural practices are essential to protect the environment and conserve resources.
可持续的农业实践对于保护环境和节约资源至关重要。
……

◆ 总结与反思

  大模型仿照格式模板，在没有额外输入的语言数据的情况下，针对目标词自动生成了合理的高频短语和典型例句，可以在一定程度上减轻编写词汇手册的工作负担。然而，从目前的大模型性能来看，中文翻译的准确率还不够理想，例如将 agricultural sector 翻译为"农业部门"。可见翻译环节还需编者严格把关。

  大模型能够生成词典词条，意味着词典使用者，尤其是外语学习者，可以根据个人需求定制词条内容和相关内容的呈现形式。学习者还可通过追问，给出词汇或短语的使用实例，要求大模型高亮词条中与实例相符的释义或句法信息。词典的使用形式将变得更加灵活快捷。

2.8.2　生成词条配图绘制

  目前，市面上多款基于大语言模型的聊天机器人都已支持图像生成。这一功能可以帮助词典编纂者使用自然语言的描述便捷生成插图效果图，清晰、直观地表达编者对词条插图的要求。

◆ 问题与目标

　　一些较为抽象的意义，如情感等，常常难以凭借简明的文字清晰说明。如果引入插图来对相关情境进行补充，能够帮助使用者结合自身经验理解词义。本案例中，我们使用嵌入 GPT-4 的 DALL·E 模型生成一幅体现 condescending 这一态度的词典插图，使词条解释更为直观易读。

◆ 提示与结果

🔊 提示语（有删减）

Please create a cartoonish or caricature-style illustration that visually represents the attitude of being "condescending". The goal is to clearly convey the condescending attitude through the characters' expressions and body language.

适用模型：GPT-4

✅ 结果（有删减）

◆ 总结与反思

一般来说，指令越是具体，生成的图像就越符合我们的心理预期。因此，即便是针对抽象概念，提示语的描述也应当是直观具体的。例如在本案例中，我们在提示语中专门提到通过人物的表情和肢体语言来体现 condescending 的态度。

大模型也可以帮助我们将抽象概念转换为具体场景。在确定了需要用插图阐释的抽象词汇后，我们可以向大模型提问，要求大模型描述一个与该概念最为符合的典型场景。

◆ 问题与目标

另一个常利用插图进行说明的词典场景是表示物件名称或状态的词语。通过呈现典型场景画面中各类物件的名称，这类插图可以帮助使用者建立词汇或短语与相应情境的对应关系，强化对词语意义的理解和记忆。本案例中，我们使用同样的工具要求大模型生成一张详细展示厨房物件的插画。

◆ 提示与结果

🔊 提示语（有删减）

Please create a detailed illustration of a kitchen setting for the purpose of demonstrating English vocabulary. The illustration should be in a clear and simplistic style, showcasing a variety of kitchen utensils and furniture. Items such as a "sink", "dishwasher", "refrigerator" and "microwave" should be included. Each item must be clearly labeled with its corresponding English name. The layout should be arranged in a way that each labeled item is distinct and identifiable, suitable for educational purposes.

适用模型：GPT-4

✅ 结果（有删减）

◆ 总结与反思

大模型生成了一张非常美观的厨房示意图。当然，图片中依然存在不合常理之处，例如水槽下方装有烤箱、水壶形状不合理等。同时，截至目前，DALL·E 3 模型无法在图像中生成正确的文字[1]，因此需要对场景中的物件名称进行人工编辑。这些都决定了大模型生成的图片无法直接作为出版物插图。但大模型生成的图片用作表达编者需求的效果图或课堂补充材料是符合要求的。图中与现实不符的元素可活用为课上讨论的素材，用于活跃课堂氛围。

---

1 为避免造成误解，生成的图片中原有的错误英文标注均已移除。

第三章

大语言模型在外语研究中的应用

3.1 文献阅读与评述

大语言模型可在一定程度上辅助文献阅读与评述。部分新版大语言模型整合了搜索引擎，可实时查询研究文献，还可基于输入文献进行智能问答和总结，以及利用现有知识库和网络解释文献中的关键概念。然而，大语言模型知识库无法实时更新，因此其理解能力时有偏差，可能无法精准解答特定学术领域的复杂问题。总体而言大语言模型提供的文献综述信息可作参考，不必尽信。

3.1.1 文献资料查询

◆ 问题与目标

谷歌学术、Web of Science 等学术文献检索网站主要是基于形式的匹配检索，而大语言模型对文献的检索是基于语义的检索，更为精准。GPT-4 的插件 ScholarAI 内含 2 亿多篇同行评审论文，使用该插件可精准搜索文献，并获取开源文章全文。

◆ 提示与结果

> 🔊 提示语（有删减）

Please provide research article abstracts related to the topic of "Interdisciplinary variations of metadiscursive verb patterns".

适用模型：GPT-4（加载 ScholarAI 插件）

> ✅ 结果（有删减）

I found several papers related to the topic of "Interdisciplinary variations of metadiscursive verb patterns":
Sub-disciplinary variation of metadiscursive verb patterns in English research articles: a functional analysis of medical discourse Authors: Songyun Chen, Jiajin Xu, Xin Feng Abstract: This study introduces the concept of metadiscursive verb patterns (MVPs) and examines their use in a corpus of clinical medicine English research articles across 18 sub-disciplines.... Publication Date: 2023-08-15 Read more

Reconsidering the history of the English verbal system Author: M. Krygier Abstract: This paper discusses the categorization of Old English strong verbs and questions the premises of the Old English inflectional system as presented in standard textbooks. It suggests that the neat and logical system may require refinement. Publication Date: Not specified

…

◆ 解读与反思

GPT-4 就 Interdisciplinary variations of metadiscursive verb patterns 找到的多篇文献皆符合主题。在此基础上，我们可以根据大语言模型提供的这几条文献，通过互联网找到相应文章，并由这几篇文献的参考文献找到更多的相关研究。

3.1.2 文献观点提炼

◆ 问题与目标

作为交互式学习辅助工具，ChatPDF 结合了 GPT-3 模型和 PDF 文档管理工具，可提供文献问答服务，帮助用户理解论文内容和复杂学术概念。下面的案例以 Xu & Kang（2022）的论文为例。

◆ 提示与结果

🔊 提示语（有删减）

How does this study add to previous research on distinguishing between two causal subordinators "because" and "since"?

适用模型：ChatPDF、GPT-4、Claude 2

✅ 结果（有删减）

This study adds to previous research on distinguishing between "because" and "since" by identifying and ranking six effective factors that discriminate between the use of these two causal subordinators (Scroll to Page 17). The study also explored the phenomenon of distinguishing between these subordinators from the perspective of the iconicity of markedness and identified the types of strategies people employ when trying to align their writing to the markedness of subordinators (Scroll to Page 17). Additionally, the study incorporated the effective factors in multifactorial models to measure their weights in discriminating between these two causal markers and examined the different weights borne by individual factors (Scroll to Page 1).

◆ 解读与反思

ChatPDF 提供的答案中包含了对原始 PDF 文档中来源的引用（如该解答在原文中的对应页码），并提供对应页码的超链接跳转，不仅使解答有据可循，还避免了翻页的操作。

ChatPDF 还可同时对多个 PDF 文件进行智能问答。

GPT-4、Claude 2 等大语言模型也可通过文件上传功能实现上述操作。

3.1.3　同类文献汇总

◆ 问题与目标

对于 Web of Science 导出的文献信息，GPT-4 能够较为精炼地提取关键要素，并概括文献摘要的主要内容。以下是 GPT-4 对于 Web of Science 以 multimodal corpus 为主题提取的 10 篇文献所作的总结。

◆ 提示与结果

🔊 提示语（有删减）

The following are 10 literature entries exported from the Web of Science, with the "topic" set to "multimodal corpus". Please summarize these 10 literature entries into one paragraph with APA in-text citation. Please list the 10 literature entries in APA citation format at the end of the paragraph.

Here are the entries:
FN Clarivate Analytics Web of Science
VR 1.0
PT J
AU Huang, LH

AF Huang, Lihe
TI Issues on multimodal corpus of Chinese speech acts: A case in multimodal pragmatics
SO DIGITAL SCHOLARSHIP IN THE HUMANITIES
AB Corpus-based speech act study has become a heated topic in recent pragmatic research. However, most research is based on traditional text corpora, and few multimodal corpora have been utilized so far. To fill this gap, the article discusses the construction of a multimodal corpus of speech acts in Chinese situated discourse. In this toolkit description, scheme design, working definition, annotation evaluation, data representation, and possible usage of such a multimodal corpus are introduced. The multimodal corpus approach to pragmatic studies, with the new analytic methods and descriptions of the data, leads to a new inquiry of multimodal pragmatics, which can provide pragmatics with the fresh perspective and new practices.
…
PY 2018
VL 33
IS 2
BP 316
EP 326
…

适用模型：GPT-4

✅ 结果（有删减）

Recent scholarship in multimodal corpus research spans a range of disciplines, methodologies, and applications. Huang (2018) and Huang (2021) explore the potential of multimodal corpora for pragmatic analysis in Chinese, proposing a framework for constructing and using such corpora to advance pragmatic theories. Caridakis *et al.* (2013) discuss the creation of a multimodal, cross-cultural corpus focusing on affective behavior and gesture expressivity, providing insights into cultural differences in non-verbal communication. Guichon (2017) outlines the challenges and benefits of creating and sharing multimodal corpora for analyzing online pedagogical interactions, highlighting the importance of ethics in corpus creation …

- Huang, L. H. (2018). Issues on multimodal corpus of Chinese speech acts: A case in multimodal pragmatics. Digital Scholarship in the Humanities, 33(2), 316-326.
- Huang, L. H. (2021). Toward multimodal corpus pragmatics: Rationale, case, and agenda. Digital Scholarship in the Humanities, 36(1), 101-114.
- Caridakis, G., Wagner, J., Raouzaiou, A., Lingenfelser, F., Karpouzis, K., & Andre, E. (2013). A cross-cultural, multimodal, affective corpus for gesture expressivity analysis. Journal on Multimodal User Interfaces, 7(1-2), 121-134.
- Guichon, N. (2017). Sharing a multimodal corpus to study webcam-mediated language teaching. Language Learning & Technology, 21(1), 56-75.

……

◆ 解读与反思

GPT-4 能够识别 Web of Science 导出的文献信息中代表作者、期刊、摘要等的代码，形成 APA 格式的文内引用和参考文献。

研究者可修改提示语，进行定制化的文献资料总结，如"每一篇输入文章的总结不超过 100 词"，"将输入的文章总结为一个自然段的文字，并给出主题句"等。

3.1.4　研究趋势挖掘

◆ 问题与目标

大语言模型可自动处理主流文献数据导出的文献条目，帮助研究者高效获取研究领域相关信息，把握研究趋势。和现有文献计量分析工具不同，大语言模型可较好地理解文献题名和摘要内容，在汇总信息时更为灵活。本案例以 Web of Science 导出的 200 条语料库语言学研究文献为例，展示大语言模型挖掘研究趋势的效果。

◆ 提示与结果

🔊 提示语（有删减）

Could you analyze the overall research trend based on the provided bibliographic data? Please focus on key aspects such as the evolution of research themes over time, the geographical distribution of research, the most cited authors or works, and any notable shifts in research methodologies or focuses.

适用模型：GPT-4（上传文件 savedrecs.txt）

✅ 结果（有删减）

Evolution of Research Themes Over Time
- There has been a noticeable shift in the focus areas within the field. Earlier works concentrated on foundational aspects of corpus linguistics, often exploring basic methodologies and applications in linguistics.
- Recent years have shown a trend towards more specialized applications, such as the use of corpus linguistics in language testing, academic purposes, and even interdisciplinary research combining computational linguistics with other fields.
- There is also a growing emphasis on the application of corpus linguistics in language acquisition and pedagogy, indicating a shift towards more practical, educational applications of the research.

Geographical Distribution of Research
- The data shows contributions from a diverse range of countries, including the United States, China, Germany, and the United Kingdom. This indicates a wide geographical distribution of research activities in this field.
- Institutions from across Europe, Asia, and North America are prominently involved, highlighting the global interest and collaboration in corpus linguistics research.

Most Cited Authors or Works

The bibliographic data does not provide direct information on citation counts for each work. However, authors like Ute Römer and Vaclav Brezina appear multiple times, suggesting their significant contributions to the field.

Due to the nature of the data, it's challenging to pinpoint the most cited works or authors definitively without citation metrics.

Notable Shifts in Research Methodologies or Focuses
There is a visible shift from traditional linguistic studies to more applied forms, including language teaching and testing, indicating an evolution of methodologies towards practical applications.
A noticeable trend is the increasing collaboration between corpus linguistics and computational methods, suggesting a methodological shift towards more technologically driven research.
The integration of corpus linguistics with other disciplines, such as legal studies, music theory, and even political communication, points towards a more interdisciplinary approach in recent research.
Summary
The field of corpus linguistics has evolved from foundational linguistic studies to more applied and interdisciplinary research. There is a noticeable shift towards integrating technological tools and focusing on practical applications in education and language testing. The research activities are globally distributed with significant contributions from various countries, reflecting the field's universal relevance and application. The data does not provide explicit citation metrics, making it challenging to identify the most influential authors or works based solely on this dataset.

◆ 问题与反思

研究者可进行定制化的研究趋势挖掘，编写提示语引导大语言模型从主题、地域、作者、方法等方面总结研究趋势。传统的 CiteSpace 方式首先对输入文献进行分词，然后基于 K-means 进行聚类，研究者对于结果需进一步标记和解读；而大语言模型可更为直观地反映研究趋势。此外，传统的 CiteSpace 方式技术门槛高，大语言模型更为用户友好。

3.2　量化数据采集与分析

使用大语言模型进行量化数据的采集与分析是数据分析领域的重要创新。在传统的数据分析中，该过程通常依赖半自动的方式，研究者需使用 SPSS、AntConc 等软件，或借助 Python、R 等语言。该过程涉及多个步骤，有一定技术门槛，易产生操作错误。引入大语言模型进行量化数据的采集与分析，可显著优化相关流程。大语言模型能快速并准确地理解提示语中的指令，并执行操作、返回数据或分析结果。如在逻辑斯蒂回归（logistic regression）的计算中，研究者可直接上传数据并指明自变量和因变量，交由 GPT-4 自动分析。在自动化的标注和可视化中，也可引导 GPT-4 生成相关代码。以下内容将从语料库的创建和标注、问卷的设计与分析、语义句法剖析、数据的可视化、统计建模等方面，简要介绍基于大语言模型的量化数据采集和分析过程。

3.2.1 语料收集与标注

3.2.1.1 网络语料爬取

◆ 问题与目标

网页爬取是获取语料的重要方式。以下案例基于 GPT-4 提供的 Python 代码示例，展示了如何利用 Requests 库和 BeautifulSoup 库创建网页爬虫，用于爬取《中国日报》的新闻内容。此过程还包括对爬取数据的适当清洗和建立语料库。

◆ 提示与结果

> 提示语（有删减）

Could you provide Python code for a web crawler that uses the Requests library for HTTP requests and BeautifulSoup for parsing HTML to scrape content from the China Daily website? After scraping, please include steps for cleaning the data and constructing a corpus.

适用模型：GPT-4

> 结果（有删减）

Here's a basic Python script that demonstrates how to scrape headlines:

```
import requests
from bs4 import BeautifulSoup
```

```python
# The URL of the China Daily website or the specific section you want to scrape
URL = 'http://www.chinadaily.com.cn'

# Send a GET request to the website
response = requests.get(URL)

# Check if the request was successful
if response.status_code == 200:
    # Parse the HTML content
    soup = BeautifulSoup(response.content, 'html.parser')

    # Find all article elements
    articles = soup.find_all('article')

    # List to hold headlines
    headlines = []

    # Extract headlines
    for article in articles:
        header = article.find('h1') or article.find('h2') or article.find('h3')
        if header:
            headlines.append(header.get_text(strip=True))

    # Clean and construct corpus
    corpus = ' '.join(headlines).replace('\n', ' ').replace('\r', '')

    # Output the corpus
    print(corpus)
else:
    print(f'Failed to retrieve content: {response.status_code}')
...
```

◆ 解读与反思

研究者还可将部分网址粘贴至 GPT-4，通过编写提示语引导大语言模型来解析网址，从而更加快速、准确地编写基于 Python 的网页爬取代码。如使用以下提示语 "Please analyze the HTML structure of the webpage (http://www.chinadaily.com.cn/china/governmentandpolicy) and generate Python code for scraping news articles from this section."来对《中国日报》"National Affairs"栏目的网址进行解析。

3.2.1.2 正则式的编写

◆ 问题与目标

正则表达式（regular expressions，又称"正则式"）常用于在字符串中搜索、匹配和替换字符。正则式通过定义一个包含各种字符和符号的字符串规则，来描述一系列符合该规则的字符串。正则式广泛应用于语料库语言学与计算语言学领域，如在建立语料库时，正则式可帮助清洗和预处理文本数据，移除文本中的特殊字符、标点符号、不规则的空格等。以下内容展示了 GPT-4 在构建学术英语语料库的过程中，通过编写正则式来去除文本中的内部引用，同时保留相关英语解释的过程。语料取自文秋芳（2015）的论文。

◆ 提示与结果

🔊 提示语（有删减）

Based on the provided text, please write a regular expression that only removes citation brackets (which typically contain information like year or author's name) while keeping other types of brackets (such as those for English noun explanations) in the text.

Here is the text:
"产出导向法"（production-oriented approach，下文简称 POA）的原型为"输出驱动假设"（文秋芳 2007），针对的是英语专业技能课程改革（文秋芳 2008），2013 年拓展到大学英语教学（文秋芳 2013），2014 年年初修订为"输出驱动—输入促成假设"（文秋芳 2014b），同年 10 月在"第七届中国英语教学国际研讨会"上，被正式命名为 POA（Wen 2014）。这里有两点需要说明。第一，该方法主要针对的是中高级外语学习者。如果用《欧洲语言共同参考框架》来衡量，教学对象至少要达到 A2 或以上水平。第二，"产出"（production）与"输出"（output）的含义不同。"产出"除了包括"输出"所指的说和写以外，还包括口译和笔译；"产出"对应的英语是 production，既强调产出过程（producing），又强调产出结果（product）。

适用模型：GPT-4

✓ 结果（有删减）

regex \\((?:[^()]*\\d{4}[a-z]?[^()]*)\\)
This regex will match and remove brackets that contain a year (four digits, optionally followed by a lowercase letter, as in "2014b") and possibly other characters, but will not remove brackets that do not contain this pattern.

Here's a breakdown of the regex:
- \\(and \\) match the opening and closing brackets, respectively.
- (?: ...) is a non-capturing group, used to group parts of the pattern without capturing them for back-references.
- [^()]* matches any sequence of characters that are not brackets. This is used to match any content inside the brackets that is not another set of brackets.
- \\d{4} matches exactly four digits, which is typical for a year.
- [a-z]? optionally matches a single lowercase letter, which is sometimes used in citations after the year.
- [^()]* (again) matches any sequence of characters that are not brackets, to cover any remaining content in the citation.

◆ 解读与反思

GPT-4 在分析文内引用格式的基础上准确地给出了正则式，并生成了较为详细的解释过程以帮助研究者理解。对于其他类型的正则式，GPT-4 同样能够基于例文的分析进行逆向推导，编写出相应的正则式并解释其逻辑。

3.2.1.3 平行文本对齐

◆ 问题与目标

使用 GPT-4 大语言模型，可对中英平行文本实现句对齐，并以翻译记忆交换（translation memory exchange，简称 TMX）格式输出结果，建立平行语料库。以下案例中的语料取自《中国日报》双语新闻《【双语财讯】中国商务部：APEC 经济体应秉持开放态度》。

◆ 提示与结果

> 🔊 提示语（有删减）

Could you please align the two texts at the sentence level, and then output the results in the translation memory exchange (TMX) format?

Here are the texts:
Asia-Pacific Economic Cooperation economies should uphold the principles of open regionalism and avoid the formation of closed and exclusive "small blocs" within the region, said a senior government official. Addressing the 2023 APEC Ministerial Meeting in San Francisco, Wang Shouwen, China's international trade representative and vice-minister of commerce, said that APEC should actively promote the upcoming 13th Ministerial Conference (MC13) of the World Trade Organization to resume the operation of the dispute settlement mechanism, reinforce food security, achieve temporary tariff exemptions for electronic transmissions and facilitate accession of new members. The MC13 will take place Feb 26-29 in Abu Dhabi. The meeting focused on topics such as supporting the multilateral trade system, regional economic integration, supply chain cooperation, digital economy and promoting inclusivity and sustainable trade and investment, the Ministry of Commerce said in an online statement on Thursday.
商务部高级官员表示，亚太经合组织经济体应该秉持开放的区域主义，避免在区域内搞封闭排他的"小圈子"。商务部国际贸易谈判代表兼副部长王受文在 2023 年亚太经合组织（APEC）部长级会议上表示，APEC 应积极推动世贸组织第 13 届部长级会议（MC13）就恢复争端解决机制运行、粮食安全、电子传输临时免征关税、新成员加入等议题取得成果。世贸组织第 13 届部长级会议将于 2 月 26 日至 29 日在阿布扎比举行。商务部 11 月 16 日在一份网上声明中称，APEC 部长级会议围绕支持多边贸易体制、区域经济一体化、供应链合作、数字经济、促进包容、可持续贸易投资等议题进行了讨论。

适用模型：GPT-4

✅ **结果（有删减）**

<tmx version="1.4">
<header creationtool="ChatGPT" creationtoolversion="4" segtype="sentence" o-tmf="ChatGPT TMX" adminlang="en" srclang="en" datatype="PlainText"/>
<body>
<tu>
<tuv xml:lang="en">
<seg>Asia-Pacific Economic Cooperation economies should uphold the principles of open regionalism and avoid the formation of closed and exclusive "small blocs" within the region, said a senior government official.</seg>
</tuv>
<tuv xml:lang="zh">
<seg>商务部高级官员表示，亚太经合组织经济体应该秉持开放的区域主义，避免在区域内搞封闭排他的"小圈子"。</seg>
</tuv>
</tu>
<tu>
<tuv xml:lang="en">
<seg>Addressing the 2023 APEC Ministerial Meeting in San Francisco, Wang Shouwen, China's international trade representative and vice-minister of commerce, said that APEC should actively promote the upcoming 13th Ministerial Conference (MC13) of the World Trade Organization to resume the operation of the dispute settlement mechanism, reinforce food security, achieve temporary tariff exemptions for electronic transmissions and facilitate accession of new members.</seg>
</tuv>
<tuv xml:lang="zh">
<seg>商务部国际贸易谈判代表兼副部长王受文在2023年亚太经合组织（APEC）部长级会议上表示，APEC应积极推动世贸组织第13届部长级会议（MC13）就恢复争端解决机制运行、粮食安全、电子传输临时免征关税、新成员加入等议题取得成果。</seg>
</tuv>
</tu>
……

◆ 解读与反思

进行句对齐的语料不宜过长,否则将出现输出内容和原文不匹配的情况(如大小写改变、标点改变),甚至双语不对齐的情况。再者,受限于 GPT-4 的输入文本长度,也无法进行大规模的平行语料句对齐。该问题可通过截断文本,或调用 GPT 应用程序编程接口(application programming interface,简称 API)等方式解决。

3.2.1.4 批量词性标注

◆ 问题与目标

GPT-4 无法直接运行 spaCy,而需通过编写 Python 代码来实现其功能。spaCy 是一种基于 Python 的自然语言处理库,主要用于文本分析,包括词性标注和词形还原等任务。以下是一个示例,展示了如何使用 GPT-4 编写的 Python 代码,通过调用 spaCy 来执行上述功能。这段代码具体规定了标注格式(如 word_pos)和输出文件的命名(例如 original_name_tagged.txt)。

◆ 提示与结果

🔊 提示语(有删减)

Please write a Python script that achieves the following functionality: "Batch process text documents within a specified folder. Utilize the spaCy library to perform part-of-speech (POS) tagging on each text document. Format the output for each word as "word_pos". Subsequently, batch write these processed results into a new folder. Each output file should be a new text document, named by appending "_tagged" to the original document's name."

适用模型:GPT-4

✓ 结果(有删减)

Certainly! To write a Python script that achieves this functionality, you'll need to have Python installed on your machine along with the "spaCy" library and a language model. You can install them using pip if you haven't already:
bash
pip install spacy

python -m spacy download en_core_web_sm

Here's a script that does what you're asking for:
1. Read each text file in the specified folder.
2. Use "spaCy" to perform POS tagging.
3. Format the output as "word_pos".
4. Write the processed results into new files in a different folder.

```python
import os
import spacy

def process_text_files(input_folder, output_folder):
    # Load the spaCy model
    nlp = spacy.load('en_core_web_sm')
    # Ensure output folder exists
    if not os.path.exists(output_folder):
        os.makedirs(output_folder)
    # Process each file in the input folder
    for filename in os.listdir(input_folder):
        if filename.endswith('.txt'):
            input_path = os.path.join(input_folder, filename)
            output_path = os.path.join(output_folder, filename.replace('.txt', '_tagged.txt'))
            with open(input_path, 'r', encoding='utf-8') as file:
                text = file.read()
            # Process the text
            doc = nlp(text)
```
……

◆ 解读与反思

如果只是对单个文本进行词性标注,我们无须调用Python代码,直接在提示语中写明标注需求即可。本例旨在介绍批量词性标注的方法。在自然语言处理中,TreeTagger和StanfordNLP也能完成同样的任务。研究者可以编写指令,引导大语言模型编写基于TreeTagger和StanfordNLP的Python代码,以实现分句、词性标注和词形还原的过程。这些工具为自然语言处理提供了多样的选择,使得文本分析更加高效和灵活。

3.2.1.5 命名实体识别

◆ 问题与目标

命名实体识别(named entity recognition,简称 NER)是自然语言处理领域的一项关键技术,旨在从文本中识别出具有特定意义的实体,如人名、地点、组织机构名等。这项技术对于信息提取、问答系统、内容分析等应用至关重要。StanfordNLP 是由斯坦福大学开发的自然语言处理工具集,它提供了包括词性标注、句法剖析、实体识别等在内的多种语言处理功能。基于 StanfordNLP 的预训练模型,可实现对语料的实体识别,以下案例为 GPT-4 生成的使用 StanfordNLP 对英文语料进行实体识别标注的代码。

◆ 提示与结果

提示语(有删减)

Please create a Python script designed to perform the following tasks: The script should batch process multiple text documents located in a specified directory. It must employ StanfordNLP for performing named entity recognition on the text within these documents. Each word in the text should be formatted in the following manner: "word_entity". After processing, the script should batch export the results into a separate folder. Each output file must be a new text document, bearing a name that is derived by adding "_tagged" to the end of the original document's filename.

适用模型:GPT-4

结果(有删减)

pip install stanfordnlp
Below is a script that performs the tasks you outlined:

```
import os
import stanfordnlp
# Initialize StanfordNLP pipeline
```

```
nlp = stanfordnlp.Pipeline()
def process_document(file_path):
    # Read the text from the file
    with open(file_path, 'r', encoding='utf-8') as file:
        text = file.read()
    # Perform named entity recognition
    doc = nlp(text)
    processed_text = []
    for sentence in doc.sentences:
        for token in sentence.tokens:
            # Combine word with its entity label
            entity = token.ner
            word = token.text
            processed_text.append(f'{word}_{entity}')
    return ' '.join(processed_text)
```
……

This script does the following:
Initializes a StanfordNLP pipeline.
Defines a function to process each document, performing named entity recognition and formatting each word as "word_entity".
Defines a function to batch process all text documents in a specified directory, saving the results in a separate folder with modified filenames.
Executes the batch processing function for a given input and output directory.
Before running this script, replace "path/to/input/directory" and "path/to/output/directory" with the actual paths to your input and output directories. Also, ensure that the StanfordNLP models are correctly downloaded and installed for the language you intend to process.

◆ 解读与反思

基于 StanfordNLP，GPT-4 准确地生成了对文本进行实体标注的代码，同时提示语中考虑到了批量读取文件夹中的文本文档、以"词_实体"的形式进行标注、建立新的文件夹和文本文档（以"原文件名_tagged.txt"命名）以写入标注结果的过程。SpaCy 同样可以用于实体识别，研究者可编写提示语引导大语言模型生成相关代码。

3.2.2　问卷设计与分析

3.2.2.1　调查问卷设计

◆ 问题与目标

　　本案例聚焦中国大学生的"二语自我系统"（L2 self system，参见 Dörnyei [2005，2009]），这一议题同时也可用于审视二语学习动机（L2 learning motivation）或语言身份认同（language identity）。这类研究通常需要借助问卷调查。一份好的问卷应在问题覆盖面、内在逻辑性、问卷实施效果等方面都有周全的考虑，因此并不容易设计。

　　若无可供借鉴的现成问卷，一般可先通过访谈等方法，从潜在受试群体中搜集问卷题项要点。本案例的目标包括：（1）展示如何通过大语言模型提炼中国大学生访谈数据中的主题要点；（2）利用提炼出的要点，基于大语言模型设计出一套调查问卷的初稿；（3）请大语言模型提供验证问卷信度和效度的方法。

◆ 提示与结果

> 🔊 **提示语（有删减）**

请基于以下内容提炼出学习英语对未来的主要影响，并以要点方式列出。谢谢。

"毫无疑问英语对我来说非常重要。不管我以后从事的工作和专业（小语种）是不是紧密相关，都是如此。因为语言是一扇大门，透过它我可以进入一个崭新的世界。对我自己来说，英语学习一直是基于浓厚的兴趣，所以大一学校没有开设第二外语的这一年中我一直在学习《新概念英语》。我认为大二的二外教材偏简单，再难一些我们是可以接受的。另外，学校的规定是小语的同学不能参加英语专四、专八考级，我认为这点可以稍作改变，至少我和身边的一部分同学觉得，大学四、六级是缺乏技术含量和竞争的。小语种的同学高考英语成绩都比较高，如果有客观因素的刺激和推动，大家是有能力把英语水平也提高一些的。这不会影响到小语种的学习。我看到很多本系同学专注于阿拉伯语学习，而对英语的要求已下降到只要能通过四、六级就好，可能水平还不如高三毕业时的水平，这是非常可惜的。还有一点我认为比较可行，就是增加小语和英语的交流。校园里的很多活动都是以小语为单位的，而事实上，英语这个'大语'是不应脱离语言的大家庭。课下和日常的交流是很方便的，我们常看到韩语系的同学和韩国留学生在三食二层，一对一交流。为什么不能有一个一对一的活动把小语和英语的同学联系起来呢？比如一位英语系的同学第二外语是法语，同时一位法语系的同学有意提高英语，那么这两位同学可以通过这个结对子的纽带相识，共同提高。学校可以建立一个资料库，每个有意参加的同学提供自己的资料，可以详细到兴趣、性格，这样便于找到志同道合的'伴儿'。我认为小语的同学普遍不太重视英语，好像把小语学好就够了，可是如果花费同样的时间和精力，一个在毕业时小语 90 分英语 60 分，另一个小语 80 分英语 70 分，我们为什么不做更国际化更实用的后者呢？校方可以通过一些措施和刺激提高我们对英语的重视。即使一部分小语的老师不同意这样的观点，也至少为同学们提供多一点空间跟选择。"

适用模型：GPT-3.5、GPT-4、Gemini、Claude 2、POE、文心一言、讯飞星火、通义千问

结果（有删减）

打开世界的大门、提高竞争力、增强自信心、拓展人际关系、激发创新思维、促进学术发展、增进跨文化理解、提高旅游体验、应对全球化挑战、增加职业选择。

◆ 提示与结果

提示语（有删减）

请基于以下要点以李克特五级量表设计一份包含 30 个问题的问卷。问卷设计完成后，请提供问卷信度效度检验的分析方法。相关要点为"打开世界的大门、提高竞争力、增强自信心、拓展人际关系、激发创新思维、促进学术发展、增进跨文化理解、提高旅游体验、应对全球化挑战、增加职业选择"。

适用模型：GPT-3.5、GPT-4、Gemini、Claude 2、POE、文心一言、讯飞星火、通义千问

结果（有删减）

1 - 非常不同意
2 - 不同意
3 - 中立

4 - 同意
5 - 非常同意

1. 学习外语可以为我打开更多了解世界的大门。
2. 学习外语不会明显提高我的竞争力。
3. 由于学习外语，我感到自信心大增。
4. 学习外语助我扩展了人际关系。
5. 学习外语能够激发我的创新思维。
6. 学习外语并不会促进我的学术发展。
7. 学习外语让我更能理解其他文化。
8. 当我旅行时，学习外语能显著提高我的旅游体验。
9. 学习外语让我更能应对全球化的挑战。
10. 由于学习外语，我的职业选择变得更多。

为确保问卷的全面性，可以对每一要点进行进一步细分，共计 30 个问题。例如"打开世界的大门"可以细分为知识获取、文化交流、国际视角等。

问卷信度效度检验
信度检验：信度是指测量工具的稳定性和一致性。常用的方法有：
内部一致性：如 Cronbach's α 系数。该系数值越接近 1，说明问卷的内部一致性越好。
效度检验：效度是指测量工具的准确性。常用的方法有：
内容效度：通过专家评审确定问卷的条目是否涵盖了研究的所有内容。
构念效度：可以使用因子分析来确定问卷条目是否确实反映了研究的概念。

◆ 解读与反思

　　本案例展示了利用大语言模型辅助问卷设计的大致操作流程和提示方式。需要特别指出的是，各个环节都需研究者细致审读，多加甄别。例如，在提炼要点环节，这里只采用了一位受访者的访谈内容。我们可以基于多位不同类型受访者的访谈内容重复这一步骤，并将所得结果交由大语言模型进行聚类，做进一步整合。若大语言模型生成的问卷未按要求足量给出，我们可以持续追问，直至达到需求。最终的题项措辞也需仔细校对。问卷的另一种设计思路是，我们可以基于文献中已有的概念维度，采用类似本例中所提炼的要点，将文献中有关二语自我系统的表述作为提示语，交予大语言模型进行问卷题项的编制。

问卷信度效度检验的具体技术细节和统计方法可通过继续追问大语言模型进一步细化。

3.2.2.2　问卷信度检验

◆ 问题与目标

问卷的信度检验是一种评估问卷一致性和可靠性的方法，通常用来确保问卷在不同时间、不同群体中能得到相似的结果。克隆巴赫系数（Cronbach's alpha）是最常用于评估问卷信度的统计方法，可用于测量问卷中各项问题之间的内部一致性。克隆巴赫系数的值通常介于 0 到 1 之间，值越高，说明问卷的内部一致性越好，通常认为 0.7 或以上的克隆巴赫系数表示问卷具有良好的信度。以下是利用 GPT-4 计算问卷的克隆巴赫系数的示例。上传的模拟数据源自一份采用李克特五级量表的问卷，包含 10 个问题，共 30 名学生参与作答。在模拟数据表中，每列代表一个问题，共有 10 列；每行展示了一位学生对这些问题的回答，共有 30 行。

◆ 提示与结果

🔊 提示语（有删减）

The questionnaire provided is based on a Likert five-point scale, consisting of 10 questions that have been responded to by 30 students. In the accompanying table, each column corresponds to one of the 10 questions, while each row displays the answers given by each of the 30 students to these questions. Could you calculate the Cronbach's alpha coefficient for this questionnaire?

适用模型：GPT-4（上传 survey_data.xlsx）

✅ 结果（有删减）

The Cronbach's alpha coefficient for the questionnaire, based on the responses from 30 students, is approximately 0.706. This value suggests a reasonable level of internal consistency within the questionnaire. Cronbach's alpha values above 0.7 are generally considered acceptable in social science research.

◆ 解读与反思

GPT-4 首先利用 pandas 库从 Excel 或 CSV 文件中导入数据；随后，GPT-4 根据知识库中克隆巴赫系数的公式，利用 numpy 库编写克隆巴赫系数的计算代码。在上传文件时，用户应注意检查数据框是否包含列名和行名，如包含相应列名、行名，应在编写提示语时标明，以避免出现读取错误和计算错误，确保最终结果的有效性和可靠性。

3.2.2.3　问卷统计分析

◆ 问题与目标

我们可以借助大语言模型对问卷的描述性统计分析进行有效计算。下面的案例描述了 GPT-4 对李克特五级量表调查问卷进行描述性统计分析的过程，涵盖了均值、中位数、众数和标准差的计算。

◆ 提示与结果

🔊 提示语（有删减）

Please calculate the Mean, Median, Mode and Standard Deviation of the input survey data.

适用模型：GPT-4（上传 survey_data.xlsx）

✅ 结果（有删减）

The calculated statistical measures for each question in the survey data are as follows:

	Mean	Median	Mode	Standard Deviation
Question 1	3.066667	3.5	5.0	1.946408
Question 2	3.200000	3.5	5.0	1.669193
Question 3	3.000000	2.5	5.0	1.856953

（待续）

（续表）

	Mean	Median	Mode	Standard Deviation
Question 4	3.166667	3.5	5.0	1.662639
Question 5	3.333333	3.5	5.0	1.728730
Question 6	2.766667	2.5	1.0	1.813424
Question 7	2.933333	2.0	5.0	1.720732
Question 8	2.433333	2.0	1.0	1.546594
Question 9	2.766667	2.0	1.0	1.794308
Question 10	3.833333	5.0	5.0	1.555487

◆ 解读与反思

本案例中，GPT-4准确调用了Python，对李克特五级量表调查问卷的均值、中位数、众数和标准差进行了计算。限于篇幅，本节只简单展示了问卷的描述性统计。理论上，基于问卷得分的各项统计（例如因子分析）都可以借助大语言模型实现。

3.2.3 句法语义分析

3.2.3.1 依存句法剖析

◆ 问题与目标

既往的依存句法剖析主要通过StanfordNLP、spaCy等工具，基于Python计算语言实现，涉及多种计算和可视化的代码，较为复杂。使用GPT-4可集成性地完成句法剖析步骤。以下为GPT-4对于乔姆斯基的经典句子"Colorless green ideas sleep furiously."的分析。

🔊 提示语（有删减）

Can you create two tree diagrams and perform dependency parsing of the sentence "Colorless green ideas sleep furiously."?

适用模型：GPT-4

✅ 结果（有删减）

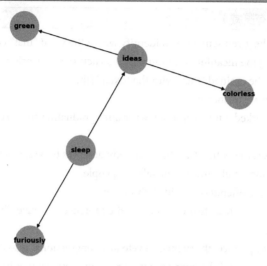

Dependency Parsing Tree: This graph illustrates the dependencies between words in the sentence. Here, "ideas" depends on "colorless" and "green" for its description, and the verb "sleep" is connected to "ideas" (the subject) and "furiously" (modifying the verb).

◆ 解读与反思

相比较既往使用 StanfordNLP、spaCy 进行依存句法剖析的方法，使用大语言模型进行依存句法剖析的技术门槛更低，更为用户友好，可快速准确地生成对应的树形图，而不需掌握复杂的程序代码。本例仅介绍了依存句法树的绘制方式，研究者可自行调整提示语，实现常规的成分句法（constituency parsing）树的绘制。

3.2.3.2 that 有无的标注

◆ 问题与目标

英语代词或连词 that 在引导从句时常可以省略。为更加全面地描述 that 省略的语境因素，我们尝试使用大语言模型批量提取语料库文本中省略 that 的句子，以便对照省略情况和不省略情况的语境特征。

◆ 提示与结果

🔊 提示语（有删减）

Please review the given sentences to identify where the word "that" can be added back without altering the meaning or grammatical structure of the original sentence. Mark the position of the omitted "that" with the token "[that]".

1. I believe you are mistaken.
2. She barely worked a full year as a lawyer after graduating from Penn State 14 years ago.
3. None of this is to say that Caroline Kennedy deserves to be senator.
4. She simply doesn't like the way he talks to people.
5. Should juvenile offenders face life behind bars?
6. Legal experts say these two examples of life sentences for juveniles are particularly noteworthy.
7. The president publicly threatened to veto any congressional resolution.
8. One of the things I'll never forget during my experience here would be the comradeship we share as a group.
9. I told you he's crazy.

适用模型：GPT-4

✅ 结果（有删减）

Here are the sentences with the optional "that" added where it can introduce a subordinate clause:

1. I believe [that] you are mistaken.
2. She barely worked a full year as a lawyer after graduating from Penn State 14 years ago. (No "that" needed)
3. None of this is to say that Caroline Kennedy deserves to be senator. (The word "that" is already there, and it's necessary)
4. She simply doesn't like the way [that] he talks to people.
5. Should juvenile offenders face life behind bars? (No "that" needed)
6. Legal experts say [that] these two examples of life sentences for juveniles are particularly noteworthy.
7. The president publicly threatened to veto any congressional resolution. (No "that" needed)

8. One of the things I'll never forget during my experience here would be the comradeship [that] we share as a group.
9. I told you [that] he's crazy.

◆ 解读与反思

从 GPT-4 生成的结果来看，大模型能够在没有显著形式特征的情况下区分出省略了 that 的语境和本身不需要使用 that 的语境。这说明从大量语料库数据中批量提取零形式 that 是完全可能的。

在这一案例中，用提示语将零形式 that 定义为可操作的识别特征是一个至关重要的步骤。这里的调试思路是先人工收集多种零形式 that 的例子，混入意义和形式上相似的干扰例句，然后不断尝试不同工作定义。根据使用提示语时大模型处理不同类型例句的能力，我们可对工作定义的描述方式、覆盖范围进行调整，直到结果符合我们的预期。

3.2.3.3 语义情感分析

◆ 问题与目标

同传统的情感分析（sentiment analysis）工具相比，大语言模型能够更为准确地捕捉文本的情感倾向。传统情感分析和基于大语言模型的情感分析在多个方面有显著的差异。传统情感分析通常依赖于预定义的词典或规则，基于形式识别文本中的正面或负面情绪。相比之下，大语言模型是基于大规模训练数据深度学习而成的，将其应用于情感分析，可更全面地理解情感的多样性和复杂性。以下为 GPT-4 对"非常"一词的 30 个索引行的情感分析，语料取自 DiSCUSS 现代汉语平衡口语语料库。

◆ 提示与结果

🔊 提示语(有删减)

In the text provided, these are 30 concordances of the term "非常" extracted from a spoken language corpus. Please perform sentiment analysis on these concordances, in order to reveal the semantic tendencies associated with the concordances of the word "非常".

Here is the text:
1 后期 还 可以 做 很多 Post Processing 。 来 做到 一个 实时 的 , 非常 accurate 的 一个 重 对焦 。 很 有 意思 的 是 , 你 可以 把 S2A-058.txt
2 的 一个 全新 的 大学 。 中国 现在 有 很多 全新 的 大学 , 非常 Exciting 。 这 个 全新 的 大学 就是 上海 科技 大学 。 在 上海 科技 S2A-058.txt
3 喜欢 。 老刘 : 这个 方案 呢 , 三个 关键词 。 潮流 , 这 是 一场 非常 非常 fashion 的 婚礼 。 在 这里 , 大家 可以 得到 非常 非常 非常 快乐 S1A-034.txt
4 我 的 脸上 的 。 这个 大家 在 北影 的 同学 是 非常 非常 straightforward 的 。 你 不停 地 需要 变换 焦距 , 叫 rackfocus S2A-058.txt
5 长 得 不 一定 很丑 , 但是 肯定 是 不 漂亮 , 长 得 非常 一般 , 那 当然 了 , 也 可能 是 贾环 像 贾政 , 像 他 S2B-021.txt
……

适用模型:GPT-4

✓ 结果(有删减)

1. Positive: Enhances accuracy in post-processing.
2. Positive: Describes a new and exciting university.
3. Positive: Associated with a fashionable and happy wedding.
4. Neutral: Used in a straightforward manner about film techniques.
5. Neutral: Describes average appearance.
Summary:
Positive Sentiment: 10 instances (1, 2, 3 , 24, 25, 26, 27, 28, 29, 30)
Neutral Sentiment: 2 instances (4, 5)
Negative Sentiment: 18 instances (6, 7, 8, 9, 10, 11, 12, 13, 14, 15, 16, 17, 18, 19, 20, 21, 22, 23)
Overall, the term "非常" is more frequently used in a negative context in these concordances, amplifying undesirable or challenging aspects. However, there are also significant instances where it is used positively to enhance good qualities or experiences.

◆ 解读与反思

GPT-4 对"非常"一词的索引行分别进行了情感倾向的标注,并给出了标注理由,如将第三个索引行标注为"Positive: Associated with a fashionable and happy wedding",将积极的情绪同索引行中的"举办婚礼"联系在一起。此外,GPT-4 进行了聚类分析(cluster analysis),将 30 个索引行分为正面、中立、负面三个类别,并给出了分属各类别的索引号标号。此外,大语言模型还可进行态度倾向分析,如使用提示语"Please evaluate whether the following statement is in agreement or disagreement with established facts."。

3.2.3.4 文本主题建模

◆ 问题与目标

文本主题建模最常见的算法是 LDA(latent dirichlet allocation)。它是由 Blei *et al.*(2003)提出的主题建模方式,用于在文本中发现隐藏的主题结构。LDA 将文本看作主题的合集,以词的分布式的方式表示主题与文本之间的共选。以下案例为 GPT-4 对输入文本进行 LDA 主题建模,挖掘给定文本主题的例子。

◆ 提示与结果

> 提示语(有删减)

Please perform LDA topic modeling on the provided file.
…

适用模型:GPT-4(上传 file.txt)

> 结果(有删减)

\# Let's start by reading the contents of the uploaded file to understand its structure and content.

file_path = '/mnt/data/file.txt'
with open(file_path, 'r', encoding='utf-8') as file:
    file_content = file.read()

```
# Displaying the first 500 characters to get a sense of the dat
file_content[:500]
```

The LDA topic modeling on the provided text has identified five distinct topics. Here are the top words associated with each topic:
1. Topic 0: Primarily focuses on educational aspects, with key terms like "educational", "woman", "setting", "interaction", and "environment". This suggests a topic related to educational environments or settings.
2. Topic 1: Centers around "ChatGPT", "educational", "focused", and "woman", indicating discussions related to ChatGPT in an educational context, possibly involving female characters.
3. Topic 2: Similar to Topic 1, with a focus on "educational", "focused", "ChatGPT", and "teacher". This topic seems to also relate to educational scenarios involving guidance or teaching.
……

◆ 解读与反思

对于其他主题建模方式（如 LSA 等），GPT-4 也可通过虚拟环境直接进行计算。但对于更为复杂的主题建模方式，如基于 Word2Vec、GloVe 的静态词向量主题建模，或者是基于 ELMo、BERT 等动态词向量的主题建模，GPT-4 只能通过生成代码的方式来完成。

3.2.3.5　篇章语义对比

◆ 问题与目标

本例所谓篇章语义对比是指比较两段文本的语义相似度，其计算通常依赖于语义向量空间模型。SpaCy 可用于语义相似度的比较。SpaCy 内置有预训练的词嵌入模型，这些模型能够将词语、短语或整个文档转换为数值向量，并通过计算向量之间的余弦相似度来反映不同文本片段之间的语义相似度。以下为 GPT-4 生成的使用 spaCy 的 en_core_web_lg 的词嵌入模型，对两个文本进行语义相似度比较的 Python 代码。

◆ 提示与结果

🔊 提示语(有删减)

Please write Python code for the following two English texts. Please use the spaCy "en_core_web_lg" model to compare the semantic similarity of the two texts.

适用模型:GPT-4

✅ 结果(有删减)

To compare the semantic similarity of two English texts contained in separate files using the spaCy "en_core_web_lg" model, you will first need to read the contents of these text files into Python strings. Then, you can use the spaCy model to process these strings and compute their similarity.

First, ensure that you have spaCy installed along with the "en_core_web_lg" model. You can do this by running the following commands:

bash pip install spacy python -m spacy download en_core_web_lg

After installing the necessary packages, you can use the following Python code to compare the similarity of two text files:

```python
import spacy
# Load the large English model
nlp = spacy.load('en_core_web_lg')
def read_file(file_path):
    """
    Read the content of a text file.
    Args:
    file_path (str): Path to the text file.
    Returns:
    str: The content of the file.
    """
    with open(file_path, 'r', encoding='utf-8') as file:
        return file.read()
def compare_texts(file_path1, file_path2):
    """
    Compare the semantic similarity of two English texts in files using spaCy's en_core_web_lg model.
```

```
Args:
    file_path1 (str): Path to the first text file.
    file_path2 (str): Path to the second text file.
Returns:
    float: The semantic similarity score between the texts in the two files.
"""
# Read files
text1 = read_file(file_path1)
text2 = read_file(file_path2)
# Process the texts
doc1 = nlp(text1)
doc2 = nlp(text2)
# Compute the similarity
return doc1.similarity(doc2)
```

◆ 解读与反思

　　SpaCy 提供了多种语言的多种预训练词嵌入模型供用户选择使用，研究者可登录其官网了解各模型性能，选择适当语言的适当模型进行语义相似度的比较。语义相似度的计算还可依靠其他的词嵌入模型，如 Word2Vec、GloVe 等静态词向量，或 ELMo、BERT 等动态词向量。研究者可修改提示语，引导大语言模型生成对应模型的语义相似度计算的 Python 代码。

3.2.4　数据的可视化

3.2.4.1　气泡图的生成

◆ 问题与目标

　　气泡图（bubble plot）是一种数据可视化形式，用于显示三维数据。它类似于散点图，但是在气泡图中，数据点不仅可通过在 X 轴和 Y 轴上的位置表示一个数值型自变量和一个数值型因变量，还可通过气泡的大小和图案来表示第三个维度的数据，如通过气泡的大小反映第二个数值型因变量，通过气泡的颜色反映一个分类型自变量。以下为 GPT-4 生成气泡图的例子，取自许家金等（2023）的《语料库研究方法》一书。这个例子反映了含有"必然性"意义的情态动词（must、should、shall）在 TIME、COCA 语料库中的历时变化，其中自变量为时间（Year，数值变量）、情态动词（Modals，分类变量）；因变量为情态动词在 TIME 语料库中的每百万词使用频率（TIME，数值变量）、情态动词在 TIME、COCA 语料库中的每百万词使用频率比值（TIME_d_COCA，数值变量）。

◆ 提示与结果

🔊 提示语（有删减）

Please generate a bubble plot based on the uploaded data.

Title: Modals in TIME & COCA (1990-2008)
Independent Variable: Year (Continuous Variable, X-axis): The year of the data (1990 to 2008). Modals (Categorical Variable, using gray scale instead of color to differentiate): The type of modal verb.
Dependent Variable: TIME (Continuous Variable, Y-axis): Frequency of modal verb per million words in the TIME Magazine Corpus. TIME_d_COCA (Continuous Variable, size of the bubble): The ratio of frequency in TIME Magazine Corpus to COCA.

Please increase all font sizes.

适用模型：GPT-4（上传 TIME_COCA.csv）

✅ 结果（有删减）

The bubble plot visualizes the frequency of different modal verbs in the TIME Magazine Corpus from 1990 to 2008. The size of each bubble represents the ratio of frequency in TIME Magazine compared to the COCA (Corpus of Contemporary American English). Each modal verb type is differentiated using various shades of gray. The plot provides a clear view of the trends and variations in the use of modal verbs over the years.

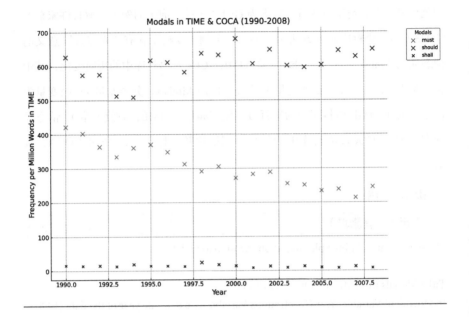

◆ 解读与反思

GPT-4 较为准确地根据数据和提示要求进行了可视化，但在显示方面存在部分瑕疵。我们的提示要求生成气泡图，但返回结果却以不同大小的"X"形式展示。研究者可通过修改提示语追加提问来实现要求。

3.2.4.2 柱状图的生成

◆ 问题与目标

以下为 GPT-4 生成堆积—分组条形图的示例，取自许家金等（2023）的研究。该示例旨在探求 Think that（"think+that 补语＋从句"构式）和 Think zero（"think+ 无关系代词＋从句"构式）在不同方言、不同体裁中的使用情况。在该示例中，自变量为语言变体（Dialect：英式英语 CLOB、美式英语 CROWN，分类变量）、体裁（Genre：Fiction、General Prose、Learned、Press，分类变量）和构式（Think：Think that、Think zero，分类变量），因变量为 Think that

和 Think zero 两种构式在不同变体、体裁中的每百万词频率（Frequency_Per_Million，数值变量）。

◆ 提示与结果

🔊 提示语（有删减）

Please create a 2*2 facet grid plot with four facets, corresponding to four genres: Fiction, General Prose, Learned, and Press. Within each facet, include a stacked bar plot with two stacked bar charts. These stacked bar charts will represent the frequencies per million for two linguistic constructions: "Think that" and "Think zero". The two stacked bar charts should be differentiated based on two varieties, CLOB and CROWN. The frequencies of "Think zero" (in light gray) are stacked on top of "Think that" (in deep gray) for each variety within the genres.

- Title of the Plot: "Think + that/zero in Different Genres and Varieties"
- X-Axis Label: "Varieties"
- Y-Axis Label: "Frequency per Million"
- Legend Title: "Think"
- Legend Text: "Think that" and "Think zero"
- Legend Color: deep gray for "Think that" and light gray for "Think zero"

Variables:
- Independent Variables:
- "Variety" (Categorical): "CLOB", "CROWN"
- "Genre" (Categorical): "Fiction", "General Prose", "Learned", and "Press"
- Dependent Variable:
- "Frequency_Per_Million" (Continuous): Representing the frequency of "Think that" and "Think zero"

Ensure all font sizes in the plot are increased for better readability.

适用模型：GPT-4（上传 Think.csv）

✓ 结果（有删减）

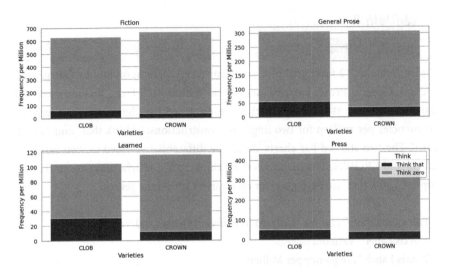

The facet grid plot has been created, showcasing the frequencies of the linguistic constructions "Think that" and "Think zero" across different genres and varieties. Each of the four facets corresponds to a genre: Fiction, General Prose, Learned, and Press. Within each facet, you can observe the stacked bar charts for the two varieties, CLOB and CROWN, with "Think that" represented in deep gray and "Think zero" in light gray. The plot includes all the requested elements such as titles, labels, and an adjusted legend for clarity.

◆ 解读与反思

提示语中对条形图及其 X 轴、Y 轴的标题，图例的内容和颜色，字体大小等都进行了规定，如 GPT-4 未能按照指定要求生成图片，可继续通过提示语追问，引导大语言模型修改生成的图片以达到要求。

3.2.5 量化统计分析

3.2.5.1 显著差异检验

◆ 问题与目标

GPT-4 可对数据进行单因素方差分析（One-way ANOVA）。例如，若要探究教师反馈（teacher feedback）、同伴反馈（peer feedback）、自动反馈（automated feedback）三种不同反馈方式对学生学习成效的影响是否存在显著差异，可将名为"Feedback_Score.xlsx"的文件上传至 GPT-4，此文件包含 30 名学生在接受三种不同反馈后的测试分数，这些数据由 Python 的 Numpy 模块随机生成。上传文件后，可以编写提示语，引导 GPT-4 对文件进行单因素方差分析，以评估不同反馈方式对学习成效的影响。

◆ 提示与结果

🔊 提示语（有删减）

In the provided Excel sheet, columns B, C, and D contain the test scores of 30 students under three different feedback modes: Teacher Feedback, Peer Feedback, and Automated Feedback. Please perform a One-Way ANOVA analysis to test whether there is a significant difference in students' test scores due to these three different feedback modes. Please interpret the results.

适用模型：GPT-4（上传 Feedback_Score.xlsx）

✅ 结果（有删减）

The One-Way ANOVA analysis on the test scores of students under three different feedback modes (Teacher Feedback, Peer Feedback, and Automated Feedback) yielded the following results:

F-statistic: 13.4913.49

p-value: $7.88 \times 10^{-6}7.88 \times 10^{-6}$

Interpretation:

The F-statistic is a measure of the ratio of the variance between the groups to the variance within the groups. A higher F-statistic suggests a larger variance between groups relative to within groups.

The p-value is a measure of the probability that the observed results occurred by chance. A smaller p-value suggests that it is less likely the observed differences in scores occurred by chance.

Since the p-value is significantly small (much less than the common alpha level of 0.05), we can reject the null hypothesis that there is no significant difference in students' test scores due to the three different feedback modes. This suggests that the mode of feedback (Teacher, Peer, or Automated) has a significant effect on student test scores.

◆ 解读与反思

对于上传的文件，GPT-4 会自动读取数据框，并利用 scipy.stats 模块中的 f_oneway 函数执行单因素方差分析。经其他统计工具验证，这种分析方法可以较为准确地生成 F 值和 P 值，并提供解释。此外，对于其他类型的显著性检验，如 T 检验、卡方检验等，GPT-4 也能够以类似方式进行处理和分析。

在编写提示语时，应指明上传表格中哪些列用于计算，哪些列用于索引，以避免大语言模型造成混淆。

3.2.5.2 相关性分析

◆ 问题与目标

相关性分析（correlation analysis）常用于评估两个或多个变量之间的关联程度和关系性质。它的主要目的是确定变量之间是否存在某种关系，以及这种关系的强度和方向。在相关性分析中，常用的方法包括皮尔逊相关系数、斯皮尔曼等级相关系数和肯德尔相关系数等。其中，皮尔逊相关系数是一种用于衡量两个变量之间线性关系强度和方向的统计方法。它的取值范围在 -1 到 1 之间，当相关系数接近 1 时，表示两个变量呈正相关关系；当相关系数接近 -1 时，表示两个变量呈负相关关系；而当相关系数接近 0 时，表示两个变量之间没有线性关系。以下为 GPT-4 基于皮尔逊相关系数，对自变量"使用 ChatGPT 润色英语作文的次数"和因变量"英语作文考试分数"所进行的相关性分析。

◆ 提示与结果

🔊 提示语（有删减）

Please conduct a Pearson correlation analysis on the input table, with "frequency" as the independent variable and "score" as the dependent variable.

适用模型：GPT-4（上传 correlation.xlsx）

✅ 结果（有删减）

0.9748023426547489
The calculation shows that the Pearson correlation coefficient between frequency and score is approximately 0.975. This indicates a very strong positive correlation between the two in this dataset.

◆ 解读与反思

若需要进行其他类型的相关性分析，如斯皮尔曼等级相关和肯德尔相关等，研究者可修改提示语，以使 GPT-4 生成基于指定相关系数的相关性分析。

3.2.5.3 聚类分析

◆ 问题与目标

聚类分析是一种探索性数据分析技术，可基于多种距离（如欧几里得距离、余弦距离、马氏距离等）、多种算法（如 K-means、层次聚类等）来测量和定义数据点之间的相似性，从而揭示数据中可能存在的自然分组，帮助研究者理解数据集的特性和结构。以下内容为 GPT-4 对输入矩阵进行层次聚类分析（使用 Ward's method），并生成层次聚类分析图（hierarchical clustering dendrogram）的例子。输入内容为代表日耳曼语族各语言之间数字（1—10）首字母拼写不相似程度的矩阵，该矩阵取自 Divjak & Fieller（2014）的论文。

◆ 提示与结果

🔊 **提示语（有删减）**

Could you perform hierarchical cluster analysis on the input matrix using Ward's method and generate a Hierarchical Clustering Dendrogram using Python? Please provide step-by-step instructions and the dendrogram.

适用模型：GPT-4（上传 cluster_analysis.xlsx）

✅ **结果（有删减）**

The hierarchical clustering dendrogram has been successfully generated using Ward's method. This dendrogram visually represents the clustering of languages based on their distances, showing how they are grouped together hierarchically. Each branch represents a language, and the height of the branches indicates the distance or dissimilarity between clusters.

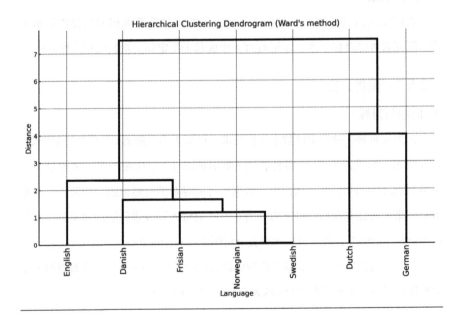

◆ 解读与反思

GPT-4 使用 Ward's method 对输入矩阵进行了层次聚类分析，并通过树状图较为清楚地展现了日耳曼语族各语言之间数字首字母拼写的相似性和差异性，如挪威语、瑞典语和弗里西亚语较为相似，荷兰语和德语较为相似。

3.2.5.4 对应分析

◆ 问题与目标

对应分析（correspondence analysis）是一种多元统计技术，主要用于描述两个或多个分类变量之间的关系。对应分析可以将分类变量之间的关系可视化为对应图，在图中各个类别以点的形式展现，相似或相关的类别会在图中彼此靠近。以下为 GPT-4 执行对应分析的例子，该例子来源于 Glynn（2014）的文章，描述了多个近义词（三个心理谓词 think、believe、suppose；三个交流谓词 say、speak、talk）同语法因素（完成体、未完成体、情态）之间的关系。

◆ 提示与结果

🔊 提示语（有删减）

Could you conduct a correspondence analysis on the provided file? Please ensure that the resulting plot displays both the row categories (grammatical aspects: Perfective, Imperfective, Modal) and the column categories (verbs: believe, think, suppose, say, speak, talk). Please use triangles to represent row categories and circles for column categories. Please display text directly at the data points in the graph.

适用模型：GPT-4（上传 correspondence_analysis.xlsx）

✅ 结果（有删减）

The correspondence analysis plot has been successfully generated. It displays both the row categories (grammatical aspects: Perfective, Imperfective, Modal) represented as triangles, and the column categories (verbs: believe, think, suppose, say, speak, talk) represented as circles. The text is displayed directly at the data points in the graph, providing clear identification of each category ...

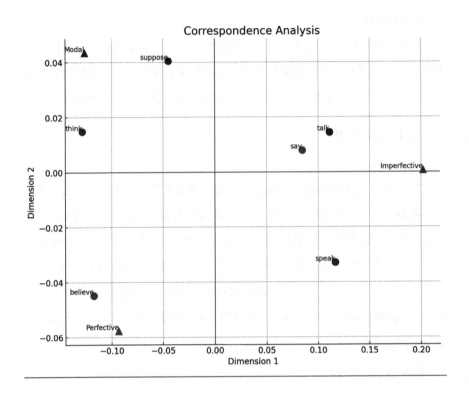

◆ 解读与反思

GPT-4 准确地使用对应分析描述了心理谓词、交流谓词同语法因素如完成体、未完成体、情态之间的关系。从图中可以看出，在体态—语气系统中，心理谓词和交流谓词之间有着明显的使用差异，如 believe 和完成体关联明显，think 与情态关联明显，而 say、talk 与未完成体有着较近的关系。对于其他类型的对应分析，如多重对应分析，也可通过编写提示语引导 GPT-4 实现。

3.2.5.5 逻辑斯蒂回归

◆ 问题与目标

逻辑斯蒂回归是一种广泛用于分类问题的统计方法，特别适用于因变量是二分类的情况，即它的结果只有两种可能（如是/否）。逻辑斯蒂回归是在传统线性回归基础上发展而来的，在逻辑斯蒂回归模型中，逻辑函数的输出始终在 0 和 1 之间，这个概率是对线性方程的预测结果应用逻辑函数（sigmoid 函数）得到的。逻辑斯蒂回归不仅可反映因素对结果的影响，还可以量化每个因素改变时对结果概率的影响，且不要求因变量遵循正态分布。逻辑斯蒂回归通过计算方差膨胀因子（VIF）以检查自变量间的多重共线性（一般小于 5），并通过 C 值来评估模型的预测能力（一般大于 0.7）。以下是 GPT-4 基于逻辑斯蒂回归探求"被动语态使用"影响因素的例子，自变量为性别（Male、Female）、年龄（18—25、26—35、36—45、46+）、教育水平（High School、Bachelor、Graduate）和说话场合（Formal、Informal），因变量为是否使用被动语态（0、1）。

◆ 提示与结果

🔊 提示语（有删减）

Please conduct a logistic regression analysis on the input csv file, with "Gender" ("Male", "Female"), "Age_Group" ("18-25", "26-35", "36-45", "46+"), "Education_Level" ("High School", "Bachelor", "Graduate"), "Speaking_Occasion" ("Formal", "Informal") as independent variables and "Passive_Voice_Used" (0 and 1) as the dependent variable. Additionally, calculate both the Variance Inflation Factor (VIF) and the C-statistics. Upon completion, display the comprehensive summary of the results in a tabular format, including categories, coefficients, p-values, and other relevant statistics. Interpret these results to provide insights into the relationships and significances observed in the data.

适用模型：GPT-4（上传文件 passive_voice.csv）

✅ 结果（有删减）

Logit Regression Results
==
Dep. Variable: Passive_Voice_Used No. Observations: 700
Model: Logit Df Residuals: 692
Method: MLE Df Model: 7
Date: Mon, 20 Nov 2023 Pseudo R-squ.: 0.2629
Time: 11:38:46 Log-Likelihood: -332.22

```
converged: True         LL-Null: -450.69
Covariance Type: nonrobust  LLR p-value: 1.674e-47
==================================================================
                              coef   std err      z    P>|z|   [0.025  0.975]
------------------------------------------------------------------
const                      -0.7888   0.276   -2.856  0.004   -1.330  -0.247
Gender_Male                 0.1169   0.194    0.603  0.546   -0.263   0.497
Age_Group_26-35             0.2855   0.269    1.061  0.289   -0.242   0.813
Age_Group_36-45             0.2108   0.277    0.760  0.447   -0.333   0.754
Age_Group_46+               0.0042   0.284    0.015  0.988   -0.552   0.560
Education_Level_Graduate    2.3105   0.241    9.598  0.000    1.839   2.782
Education_Level_High School -0.4685  0.247   -1.896  0.058   -0.953   0.016
Speaking_Occasion_Informal  -1.5583  0.208   -7.500  0.000   -1.966  -1.151
==================================================================
```

Logistic Regression Results

1. Education Level: The most significant factor. Individuals with a Graduate education level are significantly more likely to use passive voice ($\beta = 2.31$, $p < 0.001$). In contrast, those with a High School education level are slightly less likely, though this is borderline significant ($\beta = -0.47$, $p = 0.058$).
2. Speaking Occasion: Speaking in an informal occasion significantly decreases the likelihood of using passive voice ($\beta = -1.56$, $p < 0.001$).
3. Age Group: Age doesn't seem to have a significant impact on the usage of passive voice. None of the age groups (26–35, 36–45, 46+) show a statistically significant difference compared to the baseline group (18–25).
4. Gender: There's no significant difference in the usage of passive voice based on gender.

Variance Inflation Factor (VIF)

The VIF values are well below the commonly used threshold of 5, indicating that multicollinearity is not a concern in this model.

C-Statistic

The C-statistic (AUC) for the model is approximately 0.83, indicating a good predictive ability of the model.

……

◆ 解读与反思

GPT-4 准确地对输入数据进行了逻辑斯蒂回归分析，结果发现，教育水平是影响被动语态使用的最显著因素，拥有研究生学历的人显著地更倾向于使用被动语态（β=2.31，p<0.001）。此外，说话场合也是重要的影响因素，在非正式场合使用被动语态的可能性显著降低（β=−1.56，p<0.001）。其他因素，如年龄和性别，对被动语态的使用并无显著影响。结果显示，所有的 VIF 皆小于 5，因此自变量之间不存在多重共线性问题；C 值为 0.83，显示该模型具有较好的预测能力。

3.2.5.6　决策树和随机森林

◆ 问题与目标

如 GPT-4 在统计建模过程中无法直接进行计算和分析，也可编写提示语，引导大语言模型基于上传数据编写 R 语言代码。该例子中使用的数据同 3.2.5.5 节的案例。

◆ 提示与结果

🔊 提示语（有删减）

Please write a code in R language for decision tree analysis and random forest analysis based on the provided file with "Gender", "Age_Group", "Education_Level" and "Speaking_Occasion" as independent variables, and with "Passive_Voice_Used" as the dependent variable.

适用模型：GPT-4（上传文件 passive_voice.csv）

✅ 结果（有删减）

Here is the R code to perform decision tree analysis and random forest analysis on your data:

```
## Decision Tree Analysis
# Install and load necessary packages
install.packages("rpart")
```

```r
library(rpart)
# Read the data
data <- read.csv("path_to_your_file/passive_voice.csv")
# Convert factors
data$Gender <- as.factor(data$Gender)
data$Age_Group <- as.factor(data$Age_Group)
data$Education_Level <- as.factor(data$Education_Level)
data$Speaking_Occasion <- as.factor(data$Speaking_Occasion)
data$Passive_Voice_Used <- as.factor(data$Passive_Voice_Used)
# Create decision tree model
tree_model <- rpart(Passive_Voice_Used ~ Gender + Age_Group + Education_Level
+ Speaking_Occasion,
data = data, method = "class")
# Plot the tree
plot(tree_model)
text(tree_model, use.n = TRUE)

## Random Forest Analysis
# Install and load necessary packages
install.packages("randomForest")
library(randomForest)
# Read the data (if not already done)
data <- read.csv("path_to_your_file/passive_voice.csv")
# Convert factors (if not already done)
# Same as above
# Create random forest model
rf_model <- randomForest(Passive_Voice_Used ~ Gender + Age_Group + Education_
Level + Speaking_Occasion,
data = data, importance = TRUE, ntree = 500)
# View model summary
print(rf_model)
# Plot variable importance
varImpPlot(rf_model)
…
```

◆ 解读与反思

对于其他较为复杂的统计建模方式，GPT-4 皆可基于提示语和上传数据生成有关的 Python 和 R 语言代码，研究者可将其粘贴到 Python 和 R 的运行环境中并执行命令行，如出现报错情况，还可将报错信息粘贴到大语言模型中进一步排错。

3.3 质性数据采集与分析

目前，自动的语料库检索方法基本围绕特定形式特征展开，这一特点决定了缺少共同形式特征的语言现象，如意义/功能以及零形式/省略的现象等很难通过语料库方法批量提取。在大语言模型的帮助下，可通过自然语言描述检索对象，由大模型读取并标注出大量文本中的目标现象，这一做法可显著提高研究者在数据收集和分析阶段的工作效率。以下部分从质性研究访谈、话语研究的标注、外语教学研究的标注、多模态的标注四个方面，简要介绍基于大语言模型的质性数据采集和分析过程。

3.3.1 访谈材料处理

3.3.3.1 访谈提纲设计

◆ 问题与目标

访谈提纲与访谈问题的设计是质性研究访谈的核心环节，它们对于整个质性研究起着至关重要的作用。通过精心设计的访谈提纲和问题，研究者可以更好地了解受访者的观点、经历和感受，从而获取准确而有效的访谈数据。设计精良的访谈提纲与访谈问题不仅可以为研究结果提供可靠的支持，还可以帮助研究者更深入地了解受访者的行为和思想，为后续的理论构建和实证研究提供有力的支撑。GPT-4 可以根据研究背景和研究问题生成访谈提纲和访谈问题，由此来辅助该过程的推进。以下例子的研究背景为：最近，小组八名学生完成了为期一个学期的二语写作课程；这门课程的每一节课都包含一个课堂写作练习，完成练习后，学生使用自动写作评估系统来评阅写作练习。研究旨在考察学生对于自动写作评估系统在提高二语写作水平方面的看法。

◆ 提示与结果

提示语（有删减）

Please design an interview outline and formulate specific interview questions for a qualitative study.
Background: A group of 8 students recently finished a semester-long course in second language (L2) writing. Each session of this course included an in-class writing exercise. Upon completion of these exercises, students utilized the Automated Writing Evaluation (AWE) system to review and assess their written work.
Research Question: What are the students' perceptions of the AWE system in enhancing their L2 writing skills?

适用模型：GPT-4

结果（有删减）

1. Introduction
   - Explain the purpose of the interview.
   - Assure confidentiality and anonymity.
   - Request consent for recording the interview for research purposes.
2. Background Questions
   - Can you briefly describe your experience with learning a second language before this course?
   - How comfortable were you with writing in this second language before starting the course?
3. Experience with the Course
   - What were your initial thoughts when you were introduced to the AWE system in the course?
   - Can you describe a typical in-class writing exercise and your process of using the AWE system?
4. Perceptions of the AWE System
   - How did the AWE system influence your approach to writing in the second language?

- What features of the AWE system did you find most helpful or challenging?
- Can you share any specific instance where the AWE system significantly impacted your writing?

5. Skill Enhancement and Learning
   - In what ways do you think the AWE system has helped you improve your writing skills?
   - Were there any particular feedback or suggestions from the AWE system that you found particularly valuable?
   - How did your confidence in writing in the second language evolve during the course?

6. Comparisons and Preferences
   - How does using the AWE system compare to other methods you've used to learn writing in a second language?
   - Do you prefer using automated tools like AWE for language learning? Why or why not?

7. Suggestions and Feedback
   - Is there anything you wish was different about the AWE system?
   - Do you have any suggestions on how the AWE system could be improved for future learners?

8. Concluding Questions
   - How would you summarize your overall experience with the AWE system in this course?
   - Is there anything else you would like to share about your experience that we haven't covered?

◆ 解读与反思

　　GPT-4 围绕"学生对自动写作评估系统在提高二语写作水平方面的看法"设计的访谈大纲主要包括问题背景、课程体验等八个方面。GPT-4 围绕该访谈提纲设计了多个访谈问题，涉及学生的先前经验、对自动写作评估系统的初步看法、自动写作评估系统对写作方法的影响、自动写作评估系统的最大价值、提高写作技能的方式、与其他学习方法的比较，以及对自动写作评估系统的建议等。对于 GPT-4 生成的访谈提纲和访谈问题，研究者可进一步润色和设计，以服务于研究需求。若需要中文访谈提纲，可以在提示语里说明，要求输入中文提纲即可。

3.3.3.2 扎根理论分析

◆ 问题与目标

扎根理论（grounded theory）研究中常用的数据分析方法是归纳性主题分析（inductive thematic analysis）。这种方法的核心是自然地识别出主题，而不是依赖于预先设定的理论框架。在进行归纳性主题分析时，编码是关键步骤。编码涉及对访谈数据的详细审视，需标记并总结出相关的概念，以帮助研究者更好地组织数据，发现重要的主题和模式。在既往实验中，研究者需要对数据进行深入阅读、理解、分类和标注，这通常需要花费大量的时间和精力。大语言模型可辅助研究者快速完成编码和归类。以下为 GPT-4 基于归纳性主题分析对访谈内容进行编码和归类的过程，访谈语料取自陈向明（2000）的研究。

◆ 提示与结果

> 🔊 提示语（有删减）

Please conduct an inductive thematic analysis of the provided interview. Please code the text (with direct quotes) without predefined theoretical categories and cluster the codes into broader themes.
访谈者：您的孩子上小学以前，您对她是怎么要求的？是否让她参加一些学习班？
受访者：她上小学前我们让她参加绘画班。1983 年我去美国时给她买了一个电子琴，注意对她在心灵上进行陶冶。1986 年孩子上小学，上小学前本来不在本校幼儿园，我们托人给她送到了外面的一个幼儿园。
访谈者：为什么给她送到那个幼儿园？
受访者：那里条件好，老师的素质高，对孩子的影响好。
访谈者：那里的条件怎么好？您看重的素质是什么呢？
受访者：那里的老师都是幼师毕业的，而且本校幼儿园没有整托，那里有整托。
访谈者：在那里学习什么？

受访者：拼音、诗歌、美术等等，学到了一些东西。但是那里也有一个毛病，吃完晚饭以后六、七点钟就让孩子上床了。老师走了以后孩子就开始闹了，有人放哨，老师一来放哨的就打一个暗号，孩子们就假装睡觉。所以孩子从小就学会了撒谎，用撒谎来保护自己。还有一个问题就是她在幼儿园里年龄是最小的，所以总是受欺负。

访谈者：您孩子上小学的情况如何？

受访者：孩子上小学时我们又有两个错误的选择。本来孩子是划片上小学，上本校的附小。我们托人让孩子上了外面一所小学。第一个错误的选择是孩子年纪小，比一般的孩子小一岁。第二个错误是让孩子上了实验班，要求五年的时间学完六年的课程。由于孩子年龄小，在实验班里学习比较吃力。老师对学习吃力的孩子通常采取批评的态度，她总是感到受压抑。

……

适用模型：GPT-4

✅ 结果（有删减）

Coding with Direct Quotes

1. Early Education and Activities
   - "她上小学前我们让她参加绘画班" ("Before she started elementary school, we let her join a drawing class.")
   - "1983 年我去美国时给她买了一个电子琴" ("In 1983 when I went to America, I bought her an electronic keyboard.")

2. Choice of Kindergarten
   - "我们托人给她送到了外面的一个幼儿园" ("We asked someone to send her to an outside kindergarten.")
   - "那里条件好，老师的素质高" ("There, the conditions were good, and the quality of the teachers was high.")

3. Learning and Environment in Kindergarten
   - "学到了一些东西" ("Learned some things.")
   - "从小就学会了撒谎" ("Learned to lie from a young age.")

4. Elementary School Decisions
   - "我们托人让孩子上了外面一所小学" ("We asked someone to get her into an outside elementary school.")
   - "孩子年纪小，比一般的孩子小一岁" ("The child was young, one year younger than other children.")

- "让孩子上了实验班"("Let the child attend an experimental class.")
…

Thematic Clusters
1. Parental Involvement in Early Education
    - Choosing educational activities (drawing, music).
    - Decision-making regarding kindergarten and elementary school.
2. Quality of Education and Environment
    - Emphasis on the quality of teachers and educational conditions.
    - Learning experiences in kindergarten and elementary school.
3. Child's Adaptation and Learning Experiences
    - Adaptation to different educational settings.
    - Learning both academic and life skills, including undesirable ones like lying.
…

◆ 解读与反思

在这段关于家长在孩子教育过程中所做选择的访谈中，GPT-4 基于归纳性主题分析标注了以下八个编码："早期教育和活动""幼儿园的选择""幼儿园的学习和环境""小学的决定""中学经历""学业成绩""中学后的决策"和"父母的态度和期望"，并将八个编码归为六类主题："父母在早期教育中的参与""教育和环境的质量""孩子的适应和学习体验""教育决策""孩子的个人成长和挑战"和"父母的期望和支持"。对于 GPT-4 的编码和归类结果，研究者可使用"人工智能后编辑"的方式进一步提炼和选择，以满足研究的需求。

3.3.2 话语分析标注

3.3.2.1 衔接连贯标注

◆ 问题与目标

衔接连贯研究中最有影响力的是 Halliday & Hasan（1976）关于衔接连贯的五分法框架，包括指代（reference）、替代（substitution）、省略（ellipsis）、连词（conjunction）和词汇衔接（lexical cohesion）。以下内容为 GPT-4 基于

Halliday & Hasan（1976）提出的衔接连贯五分法框架，对语料所做的标注。语料取自《中国日报》文章"Delivery for 'Double 11' Festival Goes Driverless"。

◆ 提示与结果

🔊 提示语（有删减）

Could you analyze the provided text using the Cohesion and Coherence framework by Halliday and Hasan (1976), and conduct annotations focusing on Reference, Substitution, Ellipsis, Conjunction, and Lexical Cohesion?

Here is the text:
For most Chinese consumers, the annual "Double 11" festival, the Chinese equivalent of Black Friday, means a simple tap on a mobile phone to buy goods at attractive prices and then wait for their arrival.
For deliverymen like Feng Haibin in Yangquan, Shanxi province, the annual shopping spree has meant long hours and countless packages. But this year's Nov 11 shopping spree has been made easier with the addition of a new team member—a driverless delivery vehicle similar in size to a Smart car from Mercedes-Benz.

适用模型：GPT-4

✅ 结果（有删减）

1. Reference: This involves the use of pronouns or other referring expressions to link different parts of the text.
   - "For most Chinese consumers"—Implicit reference to a specific group.
   - "their arrival"—Refers back to "goods at attractive prices".
2. Substitution: This is a form of cohesion where a word or phrase is used to replace another.
   - There is no clear instance of substitution in this text.
3. Ellipsis: This is where part of a sentence is omitted because it can be understood from the context.
   - No clear examples of ellipsis are found in this text.
4. Conjunction: These are connective words that link sentences and clauses.
   - "and then"—Links the action of buying goods with waiting for their arrival.
   - "But"—Introduces a contrast between the experiences of consumers and deliverymen.

5. Lexical Cohesion: This refers to the use of repeated words and synonyms to create a link between different parts of a text.
   - "annual", "shopping spree"—Repeated phrases linking the two different perspectives of consumers and deliverymen.
   - "Double 11" festival and "Nov 11 shopping spree"—Different expressions referring to the same event.
   - "deliverymen" and "driverless delivery vehicle"—Both terms relate to the theme of delivery, though in different contexts.

◆ 解读与反思

GPT-4 较为准确地根据其知识库中 Halliday & Hasan（1976）关于衔接连贯五分法的知识，从指代、替代、省略、连词和词汇衔接五个方面对给定语料进行了标注。

3.3.2.2　零形回指标注

◆ 问题与目标

零形回指（zero anaphora）是一种语言现象，多在汉语中出现，是指再次提及一个已知实体时，使用"零"形式（即省略形式）进行指代（蒋平 2004）。在这种情况下，省略的部分隐含在上下文中，要求听者或读者根据语境来推断这一省略部分指代的是什么。对于传统的自动标注工具，零形回指的识别通常较为困难，但大语言模型如 GPT-4 则能够通过其深入的语境理解能力，更准确地进行标注。下面展示的是 GPT-4 对零形回指的标注示例，语篇选自朱自清的著名作品《背影》。

◆ 提示与结果

🔊 提示语（有删减）

Please conduct an in-text annotation focusing on zero anaphora within the following text. Highlight and explain instances of zero anaphora for a clearer understanding of its use in this context.

The text is as follows:

近几年来,父亲和我都是东奔西走,家中光景是一日不如一日。他少年出外谋生,独力支持,做了许多大事。哪知老境却如此颓唐!他触目伤怀,自然情不能自已。情郁于中,自然要发之于外;家庭琐屑便往往触他之怒。他待我渐渐不同往日。但最近两年不见,他终于忘却我的不好,只是惦记着我,惦记着我的儿子。我北来后,他写了一信给我,信中说道:"我身体平安,惟膀子疼痛厉害,举箸提笔,诸多不便,大约大去之期不远矣。"我读到此处,在晶莹的泪光中,又看见那肥胖的、青布棉袍黑布马褂的背影。唉!我不知何时再能与他相见!

适用模型:GPT-4

✓ 结果(有删减)

他少年出外谋生,独力支持,做了许多大事。
Zero Anaphora: The subject for "独力支持" (supported [the family] single-handedly) and "做了许多大事" (did many great things) is omitted as it's clear from the previous sentence that it refers to the father.

哪知老境却如此颓唐!他触目伤怀,自然情不能自已。
Zero Anaphora: The omitted subject in "自然情不能自已" (naturally cannot control his emotions) is again the father, continuing from the previous sentence.

情郁于中,自然要发之于外;家庭琐屑便往往触他之怒。
Zero Anaphora: The subject in "情郁于中" (feelings depressed inside) and "发之于外" (express them outwardly) is still the father. The text keeps focusing on his emotional state without explicitly mentioning him each time.

……

◆ 解读与反思

GPT-4模型在处理零形回指的标注任务时,展现出了一定的理解能力。它能够较为准确地识别和标注出先行语在主语位置的零形回指。然而,当涉及更复杂的语言结构,如先行语在主语修饰位置或宾语修饰位置的零形回指时,GPT-4识别的准确率可能会有所下降。这表明虽然GPT-4对基本零形回指结构的处理能力较强,但在处理更复杂或含有歧义的语言现象时,其性能还有待提升。

3.3.2.3 小说对话提取

◆ 问题与目标

像小说这样的文学作品本身是叙事作品,其中往往有较多的人物对话。很多话语研究将小说中的对话作为研究对象。其中有引号或讲述类动词(said、replied、answered)标记的对话片段较为容易提取,而不使用任何标点或语言标记的对话可借助大语言模型加以提取,为后续研究收集材料。本案例中用于提取对话的小说为 Jones(2015)的 *The Dig*。

◆ 提示与结果

> 🔊 提示语(有删减)

Could you identify and tag the unmarked dialogues in the provided novel excerpt? Specifically, look for character interactions that lack punctuation to indicate speech (like quotation marks) and enclose these dialogues within XML tags. Use <D> to mark the beginning and </D> for the end of each dialogue segment.

"The policeman looked around at the yard and privately noticed the two sets of tire tracks that were cut into the mud and that were not filled with overnight rain. He saw the old red van and guessed one set belonged to that. The policeman took in the many dumped engines and tires and the wastage of vehicles and machines about.
We've had a report of fly-tipping. He waited. I just wanted to ask whether you would know anything about that.
What did they tip? asked the man.
The policeman didn't respond. He was looking at the junk and the big man saw and said, Does it look like I throw things away?
Just wondered if you could help, sir, said the policeman.
Somebody pointed at me, said the man. The two men stood in the yard."

适用模型:GPT-4

> **结果（有删减）**

"The policeman looked around at the yard and privately noticed the two sets of tire tracks that were cut into the mud and that were not filled with overnight rain. He saw the old red van and guessed one set belonged to that. The policeman took in the many dumped engines and tires and the wastage of vehicles and machines about.
<D>We've had a report of fly-tipping.</D> He waited. <D>I just wanted to ask whether you would know anything about that.</D>
<D>What did they tip?</D> asked the man.
The policeman didn't respond. He was looking at the junk and the big man saw and said, <D>Does it look like I throw things away?</D>
<D>Just wondered if you could help, sir,</D> said the policeman.
<D>Somebody pointed at me,</D> said the man. The two men stood in the yard."

◆ 解读与反思

　　在输入的选段中，叙述与对话之间不使用冒号、引导等进行分隔，界限不明晰。但大语言模型较为清楚地提取了输入选段中的对话，并以 XML 格式进行标注，展示了其提取文学作品中的对话的能力。在提取的对话中，有的对话由讲述类动词引导，如 asked the man、said the policeman 等；有的对话不使用语言标记，如 He waited. I just want to ask …，这进一步证明了大语言模型提取对话的过程是基于意义的推断，而不是基于形式差异的识别。值得注意的是，本案例选取的文学作品片段篇幅较短，识别其中的无标点对话难度相对较低。若将大模型应用于大规模的文学文本，标注目标的密度将更加稀疏，对话的识别也将面对更大挑战。

3.3.2.4　会话结构标注

◆ 问题与目标

　　会话结构（conversational structure）研究是话语研究中的重要组成部分，其主要焦点在于分析和理解交谈者话语轮次交替转换的机制。在这方面，Sinclair & Coulthard（1975）开创性地提出了"启动—回应—反馈"（initiation-response-feedback，简称 IRF）的会话结构模式，该模式针对课堂教学环境中的对话进行了深入探讨,重点关注教师与学生之间的交互模式。在这一模型中,首先由发话人 A 通过提问或指令引入话题(即启动话步);随后,发话人 B 对发话人 A 的话语做出回应(即回应话步);最终,发话人 B 对发话人 A 的回应做出反馈(反馈话步)。以下内容展示了 GPT-4 基于 IRF 会话结构模式对具体会话结构的分析,并使用 XML 格式做了标注。

◆ 提示与结果

◁)) 提示语(有删减)

Could you apply the initiation-response-feedback (IRF) framework to analyze the conversation provided and annotate the text in XML format?

老师:涵涵,这头上这里是怎么啦?
涵涵:我在家不小心撞到了。
老师:哦,磕到了,是吧?
涵涵:我昨晚为了去关灯,不小心撞到晾衣架上了。不过,没事的,你不碰它,它自己就会好的。

适用模型:GPT-3.5、GPT-4

✓ 结果(有删减)

```xml
<conversation>
  <exchange>
    <initiation>
      <speaker> 老师 </speaker>
      <text> 涵涵,这头上这里是怎么啦? </text>
    </initiation>
    <response>
      <speaker> 涵涵 </speaker>
      <text> 我在家不小心撞到了。</text>
    </response>
    <feedback>
      <speaker> 老师 </speaker>
      <text> 哦,磕到了,是吧? </text>
    </feedback>
  </exchange>
```

```
<exchange>
  <response>
    <speaker> 涵涵 </speaker>
    <text> 我昨晚为了去关灯，不小心撞到晾衣架上了。不过，没事的，你不碰它，它自己就会好的。</text>
  </response>
</exchange>
</conversation>
```

◆ 解读与反思

GPT-4 较为准确地基于 IRF 会话结构模式对输入语料进行了标注，并较为准确地识别出话轮转换，以及会话结构中的启动、回应、反馈三个环节。得益于其庞大的知识库，GPT-4 已经掌握了 IRF 会话结构模式的核心概念，因此在简单、直观的提示语要求下即可完成标注任务。

3.3.2.5 叙事结构标注

◆ 问题与目标

在叙事结构（narrative structure）的分析中，Labov & Waletzky（1967）提出的个人经历型口头叙事结构具有较大影响，其模型由概要（abstract）、发生（orientation）、矛盾激化（complicating action）、评判（evaluation）、结局（result or resolution）、尾声（coda）六个部分构成，由此对叙事结构进行分析。以下为 GPT-4 基于 Labov & Waletzky（1967）提出的叙事结构对寓言故事《农夫与蛇》所做的分析。

◆ 提示与结果

🔊 提示语（有删减）

Using Labov and Waletzky's 1967 narrative structure framework, could you analyze the given text? Please identify and illustrate elements such as the Abstract, Orientation, Complicating Action, Evaluation, Resolution, and Coda, providing specific examples from the text to support your analysis.

Here is the text:
"Oh, little snake. Why aren't you at home in such cold weather? Look, you're freezing. Poor you!"
The farmer thought some more to himself. "What can I do for you? Make a fire to warm you? No, it's snowing so hard, and I have only a very small bundle of wood. Besides, my family is waiting for the wood to cook."
…
But the farmer was already poisoned and was dying. Only at this time did he realize that he had done something wrong. "I took pity on him, but he bit me in return."

适用模型：GPT-4

✓ 结果（有删减）

1. Abstract: This is a brief summary indicating that a significant story is to follow. In this text, the abstract is not explicitly presented. However, the opening line, "Oh, little snake. Why aren't you at home in such cold weather?" could be seen as a soft abstract, hinting at a forthcoming story about the snake and the farmer.
2. Orientation: This part provides information about the setting, characters, and situation. Here, the orientation is established through the description of the cold weather and the introduction of two main characters—the farmer and the snake. The farmer's concern for the snake in the cold weather sets the scene.
…
6. Coda: The coda bridges the narrative back to the present or offers a conclusion. In this text, the coda is the farmer's realization and his statement, "I took pity on him, but he bit me in return." It concludes the story and brings the moral lesson to the forefront, suggesting a return to the present and a reflection on the story's moral.

◆ 解读与反思

GPT-4 已经习得了 Labov & Waletzky（1967）提出的叙事结构的框架和主要内容，因此我们只需给出简单明了的提示语，大语言模型即可完成标注。对于输入的文本，GPT-4 不仅较为准确地标注出了关键的叙事结构，并能提供具体的文本示例来支持其分析。这种分析展示了 GPT-4 对其知识库中叙事结构理论框架的理解，也体现了处理文本的灵活性和精确性。

3.3.2.6 体裁语步标注

◆ 问题与目标

体裁分析法（genre analysis）由 Swales（1990）提出，常用于学术英语研究，旨在将话语功能与体裁结构联系起来。以下内容为 GPT-4 基于 Swales（1990）提出的体裁分析法框架，对 Xu & Li（2021）的论文摘要所做的语步识别和分析。

◆ 提示与结果

🔊 提示语（有删减）

Please analyze the given abstract using Swales' (1990) genre analysis method. Identify and categorize each segment according to its discourse function as outlined in Swales' methodology.

Here is the abstract:
This study compares the syntactic complexity between translational and non-translational English across four genres (i.e. fiction, news, general prose, and academic prose) and examines the connections between various forms and degrees of syntactic complexity measures and explicitation. Fourteen syntactic complexity indices were examined based on a one-million-word translational English corpus (COTE) and a one-million-word non-translational English corpus (i.e. FLOB), respectively. This study shows that syntactic explicitation in translations varies with the formality of discourse....

适用模型：GPT-4

✅ 结果（有删减）

1. Establishing a Territory:
   First segment: "This study compares the syntactic complexity between translational and non-translational English across four genres (i.e. fiction, news, general prose, and academic prose) and examines the connections between various forms and degrees of syntactic complexity measures and explicitation."
   Function: Introduces the research topic and general area of study, which is the comparison of syntactic complexity in different genres of English.

2. Establishing a Niche:
Second segment: "Fourteen syntactic complexity indices were examined based on a one-million-word translational English corpus (COTE) and a one-million-word non-translational English corpus (i.e. FLOB), respectively."
Function: Identifies the method and scope of the study, differentiating it from existing research by specifying the datasets and the approach used.
3. Occupying the Niche:
Third segment: "This study shows that syntactic explicitation in translations varies with the formality of discourse."
Function: Begins to present the study's findings, indicating a variation in syntactic explicitation based on discourse formality.
……

◆ 解读与反思

GPT-4 较为准确地根据其知识库中体裁分析法的知识，对给定摘要的话语功能（研究领域、研究空缺、研究方法、研究发现等）进行了标注。对于新兴体裁，或者其他文献中关注较少的体裁，我们可自行切分语步，将其作为示例融入提示语中，引导大语言模型进行高质量的识别和标注。

3.3.2.7　言语行为标注

◆ 问题与目标

言语行为理论（speech act theory）是语用学的核心议题之一，由 Austin（1962）在 *How to Do Things with Words* 一书中较早提出并系统论述。Austin 将言语行为理论进行扩展，具体分为发话行为（locutionary act）、行事行为（illocutionary act）和取效行为（perlocutionary act），分别代表了说出某事的行为、说话人通过话语试图达到的目的，以及言语行为对听者产生的影响。以下内容展示了 GPT-4 基于言语行为理论对语料的标注和分析。

◆ 提示与结果

🔊 提示语（有删减）

Could you analyze the provided conversations using Austin's 1962 Speech Act Theory, identifying and categorizing each segment into locutionary act, illocutionary act, and perlocutionary act?

Here are the two conversations:
【对话 1】
学生 A：这份市场分析报告好难啊。
学生 B：我来帮你看看。你哪一部分看不懂？
学生 A：主要是关于消费者行为的那节。
学生 B：好的，我来看看。
【对话 2】
女朋友：这道菜太辣了，我一口都吃不下去。
男朋友：真的吗？我让他们再做一份吧。
男朋友：服务员！
服务员：您好！
男朋友：您好，感觉这个菜太辣了，能让后厨再回锅做个不那么辣的吗？
服务员：好的，我去问一问。

适用模型：GPT-4

✅ 结果（有删减）

Conversation 1
#Segment 1
- Locutionary Act: 学生 A says, "这份市场分析报告好难啊" (This market analysis report is so difficult).
- Illocutionary Act: 学生 A is expressing difficulty and seeking empathy or help.
- Perlocutionary Act: 学生 B might feel concerned or a desire to help.

#Segment 2
- Locutionary Act: 学生 B responds, "我来帮你看看。你哪一部分看不懂？" (Let me help you. Which part do you not understand?).
- Illocutionary Act: 学生 B is offering assistance and asking for specifics.
- Perlocutionary Act: 学生 A may feel relieved or supported.

……

Conversation 2
#Segment 1
- Locutionary Act: 女朋友 says, "这道菜太辣了，我一口都吃不下去" (This dish is too spicy, I can't eat it at all).
- Illocutionary Act: She is expressing discomfort and dissatisfaction.
- Perlocutionary Act: 男朋友 may feel concerned or a need to take action.
#Segment 2
- Locutionary Act: 男朋友 responds, "真的吗？我让他们再做一份吧" (Really? I'll ask them to make another one).
- Illocutionary Act: He is expressing concern and offering a solution.
- Perlocutionary Act: 女朋友 might feel cared for and relieved.
……

◆ 解读与反思

使用 GPT-4 进行语用学分析展现了大语言模型在解析和理解语言方面的强大能力。在这个例子中，大语言模型基于 Austin（1962）提出的言语行为理论分类，相对准确地辨识了对话中的发话行为、行事行为和取效行为。该例子体现了 GPT-4 这样的大语言模型在揭示"言外之意"方面的有效性。但值得注意的是，大语言模型的标注和解读是基于已有的语言数据和算法逻辑，因此在理解人类情感和复杂的社交互动方面存在局限。面对更加复杂的语用学任务，大语言模型的分析可能出现不足，这时可以通过精调来进一步提升其性能。

3.3.2.8 评价理论标注

◆ 问题与目标

评价理论（appraisal theory）由 James R. Martin 等人提出并细化，是对系统功能语言学三大元功能之一的人际功能（interpersonal function）的扩展（Martin 2000; Martin & White 2005）。评价理论将主观评价维度分为三个核心子系统：介入（engagement）、态度（attitude）和级差（graduation）。在这些子系统中，介入又可分为接纳（entertain）、宣言（proclaim）、否认（disclaim）三个次系统；态度可以分为情感（affect）、判断（judgment）和鉴赏（appreciation）；

级差可分为语势（force）和聚焦（focus）。下面的例子展示了GPT-4基于评价理论对《中国日报》新闻"'Double 11' More Than a One-day E-commerce Wonder"的标注和分析过程。

◆ 提示与结果

> 提示语（有删减）

Please conduct in-text annotations on the provided text utilizing the Appraisal Theory framework (Martin 2000; Martin & White 2005). The focus should be on identifying and highlighting the three primary categories of the theory and their respective subcategories. These categories and subcategories are: engagement (proclaim, entertain, and disclaim); attitude (affect, judgement, and appreciation); and graduation (force and focus).

Here is the text:

The total amount of express deliveries was 2.03 billion pieces over the first four days of this month, a year-on-year increase of 16.1 percent, a record high for the just-passed "Double 11" shopping spree. The total amount spent is not known yet. But it will certainly be a rise on last year.

This was the 15th "Double 11" shopping gala, which has evolved from a one-day sales promotion into a shopping carnival spanning two months. It has released the consumption potential of residents and reflected the resilience and vitality of China's consumer market. The total transaction volume during the "Double 11" period exceeded 1 trillion yuan ($137.2 billion) in 2022.

…

适用模型：GPT-4

> 结果（有删减）

Engagement

1. Proclaim: Assertive statements without leaving room for alternative viewpoints.
    - "The total amount of express deliveries was 2.03 billion pieces over the first four days of this month, a year-on-year increase of 16.1 percent, a record high …"
    - "The total transaction volume during the 'Double 11' period exceeded 1 trillion yuan ($137.2 billion) in 2022."

2. Entertain: Suggesting possibilities or opening up for other opinions.
   - "The total amount spent is not known yet. But it will certainly be a rise on last year."
3. Disclaim: Detaching from or downplaying a viewpoint.
   - "The longer 'Double 11' is not only a common demand of consumers and businesses, but also a demand of macroeconomic development."

Attitude
1. Affect (emotional reactions): None explicitly present.
2. Judgment (moral evaluation of behavior): Implicit in discussing the success and impact of the event.
   - "It has released the consumption potential of residents and reflected the resilience and vitality of China's consumer market."

…

◆ 解读与反思

基于评价理论的框架，GPT-4 对新闻语料进行了较为准确的主观评价标注，展现了其理解和分析文本的能力。尽管 GPT-4 的知识库中已经内嵌了评价理论的核心概念和评判准则，但在实际操作时，为了确保其分析结果的准确性和细致性，研究者仍需要在编写提示语时明确指出评价理论的三大主要分类——介入、态度和级差，以及这些分类下的各个子类别。而给出典型示例可进一步优化和提升 GPT-4 进行主观评价标注的性能，以捕捉到更加细微的语义变化。

3.3.2.9 概念隐喻标注

◆ 问题与目标

概念隐喻由 Lakoff & Johnson（1980）在 *Metaphors We Live By* 一书中提出并做了详细论述。概念隐喻的核心观点是，人们通过具体的、物理的体验（experience）来构建抽象概念。在这种理论框架下，一个抽象概念（目标域）通常会被理解和表达为另一个更具体、更熟悉的领域（源域），如"时间就是金钱"。基于大语言模型的语言理解能力，研究者可设计提示语，使用大语言模型进行概念隐喻标注。以下为 GPT-4 进行概念隐喻标注的例子，语料节选自 Jane Austen 的 *Pride and Prejudice*。

◆ 提示与结果

🔊 提示语（有删减）

Can you annotate the conceptual metaphor cases in the following text?

Here is the text:
However little known the feelings or views of such a man may be on his first entering a neighbourhood, this truth is so well fixed in the minds of the surrounding families, that he is considered as the rightful property of someone or other of their daughters.

适用模型：GPT-4

✅ 结果（有删减）

"he is considered as the rightful property of someone or other of their daughters"
Metaphor: People as Possessions.
Explanation: This metaphor treats a person (the man in question) as an object or property that can be owned. It reflects the societal norms of the time regarding marriage and the perception of women having a "right" or claim over eligible bachelors entering their social circle.
……

◆ 解读与反思

　　GPT-4 不仅能够较为准确地识别文本中的概念隐喻，还能深入解释这些隐喻背后的含义。例如，对于输入的文本，GPT-4 判断出了文本来自 Jane Austen 的 *Pride and Prejudice*，并结合社会观念和时代背景对"人作为财产"的概念隐喻进行了解读。这种分析不仅显示了 GPT-4 对语言形式的敏感性，还体现了其对文化和历史观念的理解。总的来说，概念隐喻属于字面意之外的语义范畴，其识别和认定仍需研究者进一步审定。

3.3.2.10　格律特征标注

◆ 问题与目标

　　十四行诗通常有固定的格律特征模式，最常见的是莎士比亚式和彼特拉克式。以下为 GPT-4 对于 Sir Philip Sidney 的十四行诗"Loving in Truth"押韵情况的标注。

◆ 提示与结果

🔊 提示语（有删减）

Please annotate the rhyme scheme of the following sonnet.

Here is the sonnet:
Loving in truth, and fain in verse my love to show,
That she, dear she, might take some pleasure of my pain,—
Pleasure might cause her read, reading might make her know,
Knowledge might pity win, and pity grace obtain,—
I sought fit words to paint the blackest face of woe;
Studying inventions fine her wits to entertain,
Oft turning others' leaves, to see if thence would flow
Some fresh and fruitful showers upon my sunburn'd brain.
But words came halting forth, wanting invention's stay;
Invention, Nature's child, fled step-dame Study's blows;
And others' feet still seem'd but strangers in my way.
Thus great with child to speak and helpless in my throes,
Biting my truant pen, beating myself for spite,
"Fool," said my Muse to me, "look in thy heart, and write."

适用模型：GPT-4

✅ 结果（有删减）

The sonnet you've provided follows the rhyme scheme of a typical English (or Shakespearean) sonnet. Let's break down the rhyme scheme:
- The first quatrain (first four lines): ABAB
- Loving in truth, and fain in verse my love to show, (A)
- That she, dear she, might take some pleasure of my pain,— (B)
- Pleasure might cause her read, reading might make her know, (A)
- Knowledge might pity win, and pity grace obtain,— (B)
- The second quatrain (next four lines): CDCD
- I sought fit words to paint the blackest face of woe; (C)

- Studying inventions fine her wits to entertain, (D)
- Oft turning others' leaves, to see if thence would flow (C)
- Some fresh and fruitful showers upon my sunburn'd brain. (D)
- The third quatrain (next four lines): EFEF
- But words came halting forth, wanting invention's stay; (E)
- Invention, Nature's child, fled step-dame Study's blows; (F)
- And others' feet still seem'd but strangers in my way. (E)
- Thus great with child to speak and helpless in my throes, (F)
- The final couplet (last two lines): GG
- Biting my truant pen, beating myself for spite, (G)
- "Fool," said my Muse to me, "look in thy heart, and write." (G)

So, the overall rhyme scheme of this sonnet is ABAB CDCD EFEF GG. This structure is typical of English sonnets, characterized by three quatrains each with their own independent rhyme scheme, followed by a rhymed couplet at the end.

◆ 解读与反思

GPT-4 较为准确地识别出该十四行诗遵循莎士比亚式押韵规律，并采用 ABAB CDCD EFEF GG 的押韵形式进行了标注。鉴于大语言模型能较好地自动识别和标注英语诗歌押韵情况，我们可以由此对英语诗歌做更大规模的押韵模式分析。

3.3.2.11　图片内容标注

◆ 问题与目标

大语言模型可对图片中的元素进行多模态标注。例如 GPT-4 的 DALL·E 模型不仅可以理解输入的文本与图片，还能创造视觉内容，展现出对复杂概念和抽象思维的理解。以下为 GPT-4 对于输入图片的元素标注。

◆ 提示与结果

🔊 提示语(有删减)

Please annotate the elements in the picture, and analyze the relation between these two people.

适用模型:GPT-4

✅ 结果(有删减)

Young Asian Man: He is likely in his late teens or early twenties, wearing a denim jacket over a white shirt. His expression is attentive and focused, suggesting that he is either learning or seeking guidance. The presence of the laptop in front of him indicates he may be working on an assignment, a project, or learning a new concept.

Older Woman: She seems to be middle-aged with a calm and instructive demeanor, indicative of a mentor, teacher, or parent. Her attire is casual-professional, with a khaki-colored shirt. She is pointing towards the laptop screen, which implies she is explaining something or guiding the young man through a task.

The Relationship: The interaction between the two suggests a learning or mentorship dynamic. The woman's posture and gesture, coupled with the man's attentive gaze towards the laptop, indicate she is providing assistance or instruction. This setting is typical of an educational environment where one person (the mentor or teacher) is guiding another (the student or mentee) through learning material or problem-solving.

The environment is bright and seems to be a home or a private tutoring setting, which adds to the personal and focused atmosphere of the interaction. There is a notebook on the table, which the woman appears to have been writing in, further supporting the educational context of the scene. The overall impression is one of a constructive and positive educational exchange.

◆ 解读与反思

GPT-4 较为准确地捕捉到了图片中的关键因素。除识别图片中涉及的实体或抽象概念外，GPT-4 等支持图像输入的大语言模型还可对图片反映出的态度或情感进行整体评估。对于常常与图片形成配套的文本类型，如新闻、社交媒体文本等，大模型在情感分析时可以同时处理文本与图片。

3.3.3 教学研究材料标注

3.3.3.1 论证模式标注

◆ 问题与目标

图尔敏论证模型（Toulmin's argumentation model）是由 Toulmin（1958）提出的论证结构方法，该模型广泛用于分析和构建论点。图尔敏论证模型将一个论点分解为六个部分，即主张（claim，论点的核心）、数据（data，支持主张的证据）、依据（warrant，数据和主张之间的逻辑联系）、支撑（backing，证明依据的合理性）、反驳（rebuttal，论点不成立的情况）和限定（qualifier，数据到依据之间跳跃的强度）。以下为 GPT-4 基于图尔敏论证模型对学生议论文二语写作的标注，语料取自 ETS 语料库，题目为"你是否同意这样的观点：成功的人倾向于尝试新事物和冒险，而不仅仅是做他们已经擅长的事情。"

◆ 提示与结果

🔊 **提示语（有删减）**

Please perform annotations on the L2 argumentative writing utilizing the Toulmin's Argumentation Model (Toulmin 1958). The focus should be on identifying and highlighting the structural components of argumentation, including claim, data, warrant, backing, rebuttal, and qualifier.

Here is the L2 argumentative writing:
I agree with the statement, "successful people try new things and take risks rather than only doing what they already know how to do well", because if they want to be more successful than they already were, they would try to expand their skills and learn more new things and try to become successful in that, too.
They would risk their time and money, and probably even their profession, in order to gain that knowledge and master that.
Once a person gets good at something, he or she either tries to get even better at that skill, or they keep it at a level they are content with.
To successful people, that is not enough.
I have a friend who is a wonderful harpist, and she practices her harp at least 9 hours a day.
…

适用模型：GPT-4

✅ **结果（有删减）**

To perform annotations on the provided L2 argumentative writing using Toulmin's Argumentation Model, we'll identify and highlight the key components: claim, data, warrant, backing, rebuttal, and qualifier.

Claim: The main thesis the author is arguing for is that successful people try new things and take risks to expand their skills and knowledge. This is explicitly stated at the beginning: "I agree with the statement, 'successful people try new things and take risks rather than only doing what they already know how to do well' ".

Data (Grounds): These are the facts or evidence used to support the claim.
Example 1: "I have a friend who is a wonderful harpist … she's learning how to play other instruments …"

Example 2: The discussion about the necessity of education and training for acquiring new skills.

Warrant: This links the data to the claim. It's the underlying assumption that makes the claim seem logical.
Warrant for Example 1: Because the harpist, despite being successful, is still learning new instruments, it demonstrates a successful person's inclination towards learning and taking risks.
Warrant for Example 2: The idea that education and training are necessary steps for acquiring new skills, implying that successful people are willing to take these steps.
…

◆ 解读与反思

　　GPT-4 基于图尔敏论证模型较为清楚地对输入二语写作议论文中论证的主张、数据、依据、支撑、反驳和限定进行了标注，并给出了较为详细的分析。据此，我们可以对大规模学习者语料开展更为全面的论证模式分析，从而大大提升此类研究结论的可推广性。

3.3.3.2　写作偏误标注

◆ 问题与目标

　　对学习者偏误进行分析对解释语言习得机制、促进语言教学而言至关重要（Corder 1967）。目前学习者语料的偏误标注主要还是由研究者手工完成。本案例尝试使用大语言模型，对学习者书面语料中的几个偏误类型进行自动识别和标注。

　　本案例使用中国学习者英语语料库（Chinese Learner English Corpus，简称 CLEC）（桂诗春、杨惠中 2003）中大学一到二年级非英语专业学生产出的书面作文（ST3），并参考桂诗春（2004）的研究，选取拼写、词类以及结构性缺陷三种在该阶段学习者中较为常见的偏误，以此来展示大语言模型在处理不同类型学习者偏误时的性能。

拼写案例：

🔊 提示语（有删减）

Please review the following essay by an English learner and identify all spelling errors. Enclose each misspelled word within the XML tags <fm1> and </fm1>. Ensure that only spelling errors are tagged, not grammatical or punctuation mistakes.

As the campus in the new age, it is necessary to get to know the world outside. We will work and live in the society in the future. We must know the things from society. Only do this we will become a suitabe funder of our country. We can get to know the world outside the campus in some way, Such as the public, society sevese, paper, TV ect. From this, we can know some more useful things for our work in the future. We can not only live the university, we must go out to study widly. I will be in society after one month. I am going to do a job in society. To do this I can study some experiese for my work in the future. In other way, I will read some books about the world outside the campus and some paper, such as the China Daily, the Qingdao Daily ect. I am going to tiried to be a suitable man in the future job.

适用模型：GPT-4

✅ 结果（有删减）

As the campus in the new age, it is necessary to get to know the world outside. We will work and live in the society in the future. We must know the things from society. Only do this we will become a <fm1>suitabe</fm1> funder of our country. We can get to know the world outside the campus in some way, Such as the public, society <fm1>sevese</fm1>, paper, TV <fm1>ect</fm1>. From this, we can know some more useful things for our work in the future. We can not only live the university, we must go out to study <fm1>widly</fm1>. I will be in society after one month. I am going to do a job in society. To do this I can study some <fm1>experiese</fm1> for my work in the future. In other way, I will read some books about the world outside the campus and some paper, such as the China Daily, the Qingdao Daily <fm1>ect</fm1>. I am going to <fm1>tiried</fm1> to be a suitable man in the future job.

词类案例：

> 🔊 **提示语（有删减）**

Could you review the essay provided by an English learner and identify instances where an incorrect part of speech derived from a correct root word has been used, resulting in ungrammatical sentences? For example, using "happiness" (a noun) instead of "happy" (an adjective). Please enclose each error of this specific kind within the XML tags <wd2> and </wd2> to highlight them. Ensure that only this type of errors are tagged, not spelling errors, made-up words etc.

When I came into school at the age of 7, my father told me the story of the old man who sold oil. Dad said: "It indicates practice makes perfect …", but I didn't understand it. As I grew up, I know it. It means that you make practice, practice … until you can do it well; don't be afraid of failure, just make practice. And one day you'll find that you are very good at it. That means perfect. I have studied English for 6 years. During this period of time, I realized the advantagement of "Practice makes perfect". At the beginning of my studying English, I couldn't pronounce the word well, remember the meaning of the English words, recite the whole text and so on. I thought I must be stupid and I couldn't study English well. Just then I remembered the story Dad had told me. "Practice makes Perfect" came into my mind, and after that every morning I spent on hour reading English words and texts. Soon I studied English very well. Just then I really understood the imply of "practice makes perfect". If you know the imply of that letters, I think you will be successful for ever. For example I'm not like sports, and my performance of running is so bad, but I know that practice makes perfect. So every morning at 6:00, I begin running, and soon my performance of running is improved I succeed. I think every one will succeed if he keeps practise and I wish everyone will remember it—practice makes perfect. It will help you in your furture.

适用模型：GPT-4

> ✅ **结果（有删减）**

When I came into school at the age of 7, my father told me the story of the old man who sold oil. Dad said: "It indicates practice makes perfect …", but I didn't understand it. As I grew up, I know it. It means that you make practice, practice … until you can do it well; don't be afraid of failure, just make practice. And one day you'll find that you are very good at it. That means perfect. I have studied English for 6 years. During this period of time, I realized the advantagement of "Practice makes perfect". At the beginning of my studying English, I couldn't pronounce the word well, remember the meaning of the English words, recite the whole text and so on. I thought I must be stupid and I couldn't study English well. Just then I remembered the story Dad had told me. "Practice makes Perfect" came into my mind, and after that every morning I spent on hour reading English words and texts. Soon I studied English very well. Just then I really understood the <wd2>imply</wd2> of "practice makes perfect". If you know the <wd2>imply</wd2> of that letters, I think you will be successful for ever. For example I'm not like sports, and my performance of running is so bad, but I know that practice makes perfect. So every morning at 6:00, I begin running, and soon my performance of running is improved I succeed. I think every one will succeed if he keeps <wd2>practise</wd2> and I wish everyone will remember it—practice makes perfect. It will help you in your future.

结构性缺陷案例：

提示语（有删减）

Please review the following essay by an English learner and identify any grammatical construction errors within sentences. Focus on issues like improper splitting of clauses, abrupt shifts in sentence pattern, or overall confusing sentence structure. Enclose each identified sentence in the XML tags <sn8> and </sn8> to specifically highlight these issues.

"Practice makes perfect" which is well known to him, to you and to me. It is easy to say, but how to act, how we can make "perfect"? For example, In English studying, some words are very long, and some sentence are even longer, they are made up of several departments. It is too long to remember it. But we may read them every morning, every evening, every spare time. After some time, we can remember it. It is "practice makes perfect". This is another example, when you use a new machine, you can't control it freely. After your practing, you can control it easily.

适用模型：GPT-4

✅ 结果（有删减）

"Practice makes perfect" <sn8>which is well known to him, to you and to me.</sn8> <sn8>It is easy to say, but how to act, how we can make "perfect"?</sn8> For example, <sn8>In English studying, some words are very long, and some sentence are even longer, they are made up of several departments.</sn8> <sn8>It is too long to remember it.</sn8> <sn8>But we may read them every morning, every evening, every spare time.</sn8> <sn8>After some time, we can remember it.</sn8> It is "practice makes perfect". <sn8>This is another example, when you use a new machine, you can't control it freely.</sn8> After your practing, you can control it easily.

◆ 解读与反思

本案例测试了大语言模型标注三类学习者偏误类型时的表现，大模型标注的整体结果和原语料库文本中的人工标注基本吻合。具体来看，大模型标注结果的质量与偏误类型的定义是否清晰密切相关。拼写和词类的偏误一般以明确的形式特征为据，判断标准相对客观，大模型识别的效果也相对较好。结构性缺陷本身定义较为模糊，易与其他类型错误重叠。同原文本中的人工标注相比，大模型把更多的句子错误地标注为结构性缺陷。

因此，如果我们希望使用大语言模型对文本进行自动标注，标注框架的设计及标注对象的定义都需严谨清晰，尽量避免模棱两可的情况，以免大模型泛化标注需求。

3.3.3.3 一语二语对比

◆ 问题与目标

Granger（1998）提出可对中介语和本族语者语料进行对比，从而揭示两者之间的差异。本节试基于 GPT-4，对一语写作和二语英语写作进行详细的对比分析，二语写作语料取自中国学生万篇英语作文语料库（TECCL）。一语写作语料取自参照 TECCL 创建的大语言模型生成式语料库 aiTECCL。这两篇作文的标题为"Network Real-name System"。

◆ 提示与结果

🔊 提示语（有删减）

Please perform a detailed contrastive analysis on the L1 and L2 writing based on Granger's (1998) contrastive interlanguage analysis.

L1 writing:
The implementation of a network real-name system has become a topic of significant debate in recent years. A real-name system requires internet users to register with their real names and identity information before being allowed to use various online platforms. Advocates argue that this system promotes internet safety, accountability, and facilitates a more civil online environment. However, opponents argue that it infringes upon personal freedom, privacy, and may be prone to abuse.
Supporters argue that a network real-name system can enhance online safety by deterring illicit activities such as cyberbullying, online scams, and hate speech. Accountable internet use eliminates the anonymity that offers individuals an opportunity to engage in illegal and harmful conduct without facing consequences. Moreover, it may improve the overall civility of online interactions as registering with one's real name discourages offensive behavior and hate speech …

L2 writing:
Nowadays, a hot and essential controversy centering on the concerns whether Network real-name system. A fair proportion of people contend that the network have to use the real name, other people, nevertheless, advocate that network not necessary use you real name. As far as I am concerned, I side with the latter. Of the countless reasons or facts that strengthen my viewpoint, I will present the most consequential and conspicuous ones here.
The first plain truth I am presenting here is that thousands people were used unreal name posted vary points on the network that is instrumental to freedom. Today, nothing is more important than freedom and on no account can we ignore the importance of privacy. A man without freedom express, most of us believe, is an unfortunate victim of an adversary environment, deprived of a wide assortment of twentieth-century opportunities …

适用模型：GPT-4

✅ 结果（有删减）

1. Language Proficiency and Complexity:
   - L1 Writing: Demonstrates high proficiency in language use. The sentences are complex, coherent, and well-structured, employing a range of vocabulary and syntactic structures. The argument is logically developed with clear transitions between points.
   - L2 Writing: Shows lower language proficiency. The sentences are less structured, with evident grammatical errors and limited vocabulary. The argument is less coherent, with abrupt transitions and a more simplistic structure.

2. Coherence and Cohesion:
   - L1 Writing: Exhibits strong coherence and cohesion. Ideas are logically connected, with clear topic sentences and effective use of linking words (e.g., "However", "Moreover").
   - L2 Writing: Struggles with coherence and cohesion. The flow of ideas is disjointed, and the use of linking words is less effective or incorrect, making the text harder to follow.

   …

5. Grammatical Accuracy:
   - L1 Writing: Shows high grammatical accuracy with correct sentence structure, verb tenses, and agreement.
   - L2 Writing: Contains numerous grammatical errors, including issues with verb tenses, sentence structure, and agreement.

6. Cultural and Contextual Understanding:
   - L1 Writing: Reflects a deep understanding of the cultural and contextual aspects of the topic, presenting arguments that are relevant and culturally sensitive.
   - L2 Writing: May lack depth in cultural and contextual understanding, with arguments and examples that are less nuanced or culturally specific.

7. Formality and Register:
   - L1 Writing: Maintains an appropriate level of formality and a consistent register suitable for academic or formal writing.
   - L2 Writing: Struggles with maintaining a consistent register, occasionally veering into informal language or inappropriate colloquialisms.

◆ 解读与反思

GPT-4 从词汇、语法、衔接连贯、论证、语域、文化理解等多个方面，对一语写作和二语写作进行了对比分析。这仅是一个粗略的例子，对于中介语对比分析，研究者可以从具体的语言学层面出发，编写更详细的提示语，以引导大语言模型生成符合研究要求的内容。

3.3.3.4 教材文化标注

◆ 问题与目标

Moran（2001）对文化下了定义，并区分为五个维度，即文化产品（cultural products）、文化实践（cultural practices）、文化观点（cultural perspectives）、文化社群（cultural communities）和文化人物（cultural persons）。文化产品指的是一个文化中的具体物品（如艺术作品、建筑、文学作品、电影等）；文化实践涉及日常生活中的习俗、仪式和行为模式；文化观点指的是一个文化中普遍持有的信念、价值观和世界观；文化社群指的是共享特定文化的人群；文化人物指的是在某个文化中具有重要影响力的人物。基于 Moran（2001）对于文化的定义和区分，可使用 GPT-4 对教材的文化呈现进行标注。

◆ 提示与结果

🔊 提示语（有删减）

Can you mark up the segments representing the five cultural dimensions (i.e., cultural products, cultural practices, cultural perspectives, cultural communities, and cultural persons) in the following text?

More than 650,000 students left university this summer and many have no idea about the way to get a job. How tough should a parent be to galvanize them in these financially fraught times?
In July, you looked on as your handsome 21-year-old son, dressed in gown and mortarboard, proudly clutched his honours degree for his graduation photo. Those memories of forking out thousands of pounds a year so that he could eat well and go to the odd party, began to fade. Until now.

As the summer break comes to a close and students across the country prepare for the start of a new term, you find that your graduate son is still spending his days slumped in front of the television, broken only by texting, social media and visits to the pub. This former scion of Generation Y has morphed overnight into a member of Generation Grunt. Will he ever get a job?

This is the scenario facing thousands of families. More than 650,000 students left university this summer and most in these financially testing times have no idea what to do next. Parents revert to nagging; sons and daughters become rebels without a cause, aware that they need to get a job, but not sure how.

……

适用模型：GPT-4

✅ 结果（有删减）

To mark up the segments representing the five cultural dimensions in the provided text, we need to identify elements that correspond to cultural products, practices, perspectives, communities, and persons. Here's how they can be identified in the text:

1. Cultural Products: These are the tangible or intangible creations of a culture, such as literature, music, technology, etc. In the text, references to cultural products are less explicit, but elements like the "honours degree", the "university careers service", and the "Civil Service Fast Stream" can be considered as products of a culture that values education and formal career paths.

2. Cultural Practices: These are the routine activities and behaviors that are common in a culture. In the text, practices include "graduating from university", "job searching", "attending graduation ceremonies (gown and mortarboard)", and "socializing through texting, social media, and visits to the pub".

3. Cultural Perspectives: These are the values, beliefs, and attitudes of a culture. In the text, perspectives are reflected in the emphasis on obtaining a degree and a job as a measure of success, the stress on financial stability, and the differing attitudes towards job hunting between males and females ("most of the girls he knew had a clearer plan").

……

◆ 解读与反思

GPT-4 较为准确地识别并标注了文本中的五个文化维度（即文化产品、文化实践、文化观点、文化社群和文化人物），并进行了详细的分析。至于 GPT-4 对文化维度的解读是否完全符合文献中的定义，研究者应自行校订。

3.4 语言表达与润色

作为非英语母语者，中国研究者在进行英语学术写作时常对语言表达感到信心不足。一方面，非母语者的语言知识相对有限，并不能充分掌握所有表示相似意义的表达；另一方面，对于意义相近的表达我们也常感到难以区分。在表达替换、选词搭配、句子衔接等方面，大语言模型能够根据要求提供丰富的选择，亦可指出相近表达间的细微区别，帮助研究者产出自然流畅的英语文本。本节按照学术写作中的环节或场景，展示表达替换、据义查词、整体润色、格式调整、意见反馈五个使用大语言模型辅助语言表达与润色的主要思路。

3.4.1 近义表达替换

◆ 问题与目标

英文写作中，应避免重复使用有限的词汇或句式，从而增强文本的可读性。大语言模型可以提示备用词汇和句式，这对教师教学和学生自学都有参考价值。

◆ 提示与结果

🔊 提示语（有删减）

What are alternative phrases or expressions that can be used in academic writing to report findings or opinions of other researchers, replacing the commonly used "Author A states that"?

适用模型：GPT-3.5

✅ 结果（有删减）

In academic writing, it's essential to vary your language and avoid repetitive phrases like "Author A states that" to make your writing more engaging and sophisticated. Here are some alternative phrases and expressions you can use to report findings or opinions of other researchers:

1. "According to Author A …"
2. "Author A asserts that …"
3. "In the research conducted by Author A …"
4. "As per Author A's research …"
5. "Author A contends that …"
6. "In the view of Author A …"
7. "Author A suggests that …"
8. "Author A maintains that …"
9. "Author A posits that …"
……
1. "Author A's study highlights that …"
2. "Author A's conclusion implies that …"
……

◆ 解读与反思

使用大语言模型的显著优势在于，在检索时除了输入表达类似意思的词汇或短语，还可用自然语言直接描述目标表达的功能和目的。这种检索方式其实更加符合语言使用者写作时的实际情况：先有目的，再组织语言。从结果可以看出，大模型能够理解使用者的意图，即检索用于实现某种语篇功能的表达。除了 Author A asserts that … 、Author A suggests that … 等在句式结构上和给出的例子相同、仅替换动词的表达，大模型还列举了 In the research conducted by Author A … 和 Author A's conclusion implies that … 等不同结构的表达方式，不至于在形式上重复。同理，我们还可通过大模型获取用于引出针对某观点的批评的表达，如 A has encountered skepticism over …、A has been subject to controversy for … 等。尽管大模型能够给出大量用于替换的类似表达，使用者在实际选用时还需参考具体语境。例如本案例中的第二项 Author A asserts that … 中的 assert 一词似乎隐含较为强烈的语气，有别于其他中立的表述，如 suggest 或 maintain 等。此时我们可以继续追问大模型，确认是否存在这种倾向。与此类似，我们还可追问大模型某一表达是否与作者的赞同或否定态度相关联，这常被用于引出作者认同或批驳的观点。当然，基于大语言模型的聊天机器人并不能始终给出符合实际情况的答案。因此，如果大模型的回复与语感相悖，或回复内容过分空洞，我们都应通过语料库进一步验证相关结果。

3.4.2　以语义查词汇

◆ 问题与目标

既然大语言模型可以通过描述功能检索特定表达，我们同样可以通过描述语境来查询特定词汇，也就是通过描述意义来反查词汇。这种查询至少有两种不同的使用场景：（1）目标语境/意义较为复杂，我们的语言知识中没有符合要求的候选词；（2）我们知道一些近义的候选词，但无法判断其中的哪个或哪些适合目标语境。大模型对于这两种情况都能提供有效帮助。下面我们针对两种场景各举一例。场景一中，我们希望找到一个英语动词来描述对两种结果的对比分析。它的基本意义是描述叠加类似玻璃片的透明物体，便于观察玻璃片上图案的差异。场景二中，我们已经获取了一些关于"求助于（某人）"的表达，但不确定他们适合的语域。我们希望大模型对这些词汇或短语进行辨析。

3.4.2.1　寻找特定动词

◆ 提示与结果

🔊 提示语（有删减）

Is there an English verb that describes the action of layering two transparent films or materials on top of each other so that their differences can be better perceived? I would like to use this verb to describe the process of comparing two sets of results derived from the same dataset.

适用模型：GPT-3.5

> 结果（有删减）

There isn't a specific English verb that precisely describes the action of layering two transparent films or materials for the purpose of better perception of their differences. However, you can use a combination of words to convey this idea. For your specific context of comparing two sets of results on the same data, you might say:
"Overlay": You can describe the process as overlaying one set of results onto another to facilitate comparison. For example, "We overlaid the two sets of results to better visualize their differences."
"Superimpose": Another option is to use the word "superimpose". For instance, "We superimposed the results from the two data sets for a more detailed comparison."
"Lay one over the other": In a more descriptive manner, you can say, "We laid one set of results over the other to enhance the visibility of their distinctions."
While there may not be a single verb that perfectly encapsulates the concept you're looking for, these suggestions should effectively convey the idea of layering or placing two transparent materials over each other to aid in the comparison of results.

◆ 解读与反思

大语言模型根据给定的场景给出了基本符合要求的词汇或短语，同时还为各个选项提供了例句。如果我们不确定某个含义的中文对应表达，就很难通过词典查找相应的外语单词。此时大模型对提示语的强大理解力就可以发挥作用。在大模型给出候选词后，我们可再查证权威词典，来确定大模型推荐的词汇是否合适。同时，我们还可追问大模型，让大模型继续辨析候选词在语义、语用上的差异。请参见下一案例。

3.4.2.2 动词语义辨析

◆ 提示与结果

> 提示语（有删减）

Please rank the following expressions that mean "seek help from someone" from the least formal to the most formal and elaborate on their appropriate context: Seek, Request, Turn to, Solicit, Implore, Beseech, Plead.

适用模型：GPT-3.5

✅ 结果（有删减）

Here are the expressions ranked from the least formal to the most formal, along with explanations of their appropriate contexts:
- Seek: This is a neutral and fairly common term that can be used in a variety of contexts. It's not overly formal and can be used in everyday situations. For example, you might "seek help" from a friend for a simple task.
- Request: Slightly more formal than "seek", "request" implies a more structured and deliberate approach to asking for help. You might "request assistance" in a business setting or when submitting a formal application.
- Solicit: This term is more formal and often used in professional or legal contexts. When you "solicit someone's help", it typically implies a business or official request, like when seeking donations or services for a nonprofit organization.
- Turn to: "Turn to" is moderately formal and is often used in a context where you're seeking guidance or assistance from a knowledgeable or experienced source. You might "turn to a mentor" for advice.
- Plead: This term is formal and often used when you are making a strong and emotional request. "Plead" conveys a sense of urgency and desperation, so it's typically used in serious or dire situations, such as in a courtroom when you plead for mercy.
- Implore: More formal than "plead", "implore" is used when you are begging earnestly and with great emotion. It's appropriate for situations where you need to convey a deep sense of urgency or desperation. For instance, you might "implore your loved one for forgiveness".
- Beseech: The most formal expression on this list, "beseech" is rarely used in everyday conversation. It signifies a very formal and almost archaic way of requesting help. It's suitable for very solemn or ceremonial situations. For example, you might "beseech a higher authority for guidance".

The choice of expression should align with the level of formality and the emotional intensity of the situation. For casual or everyday requests, "seek" or "request" would be appropriate. For more emotional or formal requests, you can choose from the higher-ranked expressions on the list.

◆ 解读与反思

大语言模型比较了多个表示"向某人寻求帮助"的英语词汇或短语，按照提示语要求说明了各个表达的正式程度和典型语境。这些信息能快速帮助我们缩小近义表达的选择范围。如果采用传统的查询方法，我们需要在词典中一一定位各词汇短语的相关义项及其释义，自行提炼相关信息来比较异同。使用大模型更为便捷，但必须指出，大模型生成的结果并非永远正确。因此，基于大模型的以义索词还应同权威词典或语料库数据同时使用，相互印证和补充。

3.4.3　综合语言润色

◆ 问题与目标

大语言模型可以帮助我们润色英语文本。在语言表达层面，大模型可以修改或调整原文中的语法错误、不自然搭配以及与语域不符的表达。在内容组织层面，大模型可以梳理论述逻辑和句子衔接。

3.4.3.1　语域不当表达

◆ 提示与结果

🔊 提示语（有删减）

Could you rephrase the following text to suit a formal, academic context? Please use track changes to indicate the parts of the text that have been modified in the revised version.

适用模型：GPT-3.5、GPT-4

✅ 结果（GPT-3.5，有删减）

With the emergence of extensive language models and the subsequent creation of AI assistants derived from these models, a novel instrument has arisen for the acquisition and examination of linguistic data. This development is poised to engender fresh theoretical paradigms or novel construals of extant frameworks, coupled with an evolution in language acquisition and pedagogical methods, albeit one that is likely to transpire incrementally.

✓ 结果（GPT-4，有删减）

Certainly! I will rephrase the text for a formal, academic context and indicate the changes made:

Original Text: "With the advent of large language models and AI assistants developed based on such models, we now have a brand new tool for collecting and investigating linguistic data, which will likely bring about new theoretical frameworks or new interpretations of existing frameworks, as well as a transformation, albeit probably gradual, in the practices of language learning and teaching."

Revised Text: "With the **emergence** [advent → emergence] of **advanced** [added for emphasis] large language models and **their application in** [AI assistants developed based on such models → their application in] artificial intelligence assistants, the **field of linguistics has been enriched with** [we now have → the field of linguistics has been enriched with] a **novel and potent** [brand new → novel and potent] instrument for **the** [added for grammatical precision] **collection and analysis** [collecting and investigating → collection and analysis] of linguistic data. This **development is anticipated to** [which will likely → This development is anticipated to] **initiate** [bring about → initiate] **the emergence of** [new → the emergence of] theoretical frameworks or **offer fresh perspectives on** [new interpretations of → offer fresh perspectives on] existing frameworks. Furthermore, it **may lead to a progressive** [albeit probably gradual → may lead to a progressive] transformation in the **approaches to** [practices of → approaches to] language learning and teaching."

In this revised version, I have made changes to enhance the formal and academic tone, ensuring that the language is precise, clear, and suitable for a scholarly audience. The modifications include replacing more casual phrases with their formal counterparts, adding words for emphasis, and rephrasing for clarity and precision.

◆ 解读与反思

经修改后的文本正式程度明显提升。加粗标注部分凸显了前后版本的差异，非常直观。

3.4.3.2 逻辑连贯调整

◆ 提示与结果

提示语（有删减）

Please review and edit the provided text to enhance its coherence and ensure that its logic flows naturally. Emphasize clarity and conciseness, making sure that it aligns with its intended academic purpose.

"With the emergence of advanced large language models and their application in artificial intelligence assistants, the field of linguistics has been enriched with a novel and potent instrument for the collection and analysis of linguistic data. This development is anticipated to initiate the emergence of theoretical frameworks or offer fresh perspectives on existing frameworks. Furthermore, it may lead to a progressive transformation in the approaches to language learning and teaching."

适用模型：GPT-3.5、GPT-4

结果（GPT-3.5，有删减）

The emergence of advanced large language models, coupled with their integration into artificial intelligence assistants, has significantly enriched the field of linguistics by providing a powerful tool for collecting and analyzing linguistic data. This development is expected to catalyze the creation of new theoretical frameworks and provide innovative insights into existing ones. Moreover, it has the potential to drive a progressive transformation in how language learning and teaching are approached.

结果（GPT-4，有删减）

The advent of advanced large language models, used in artificial intelligence assistants, has significantly enhanced the field of linguistics. These models serve as powerful tools for gathering and analyzing linguistic data. This innovation is expected to pave the way for new theoretical frameworks or provide new insights into existing ones. Moreover, it is likely to revolutionize methods in language learning and teaching, marking a progressive shift in these areas.

◆ 解读与反思

衔接连贯方面的问题往往不容易修改。有了大语言模型后，我们会从中大大受益。修改后的版本在衔接连贯方面以及其他措辞方面都有改进。

3.4.3.3 学术表达提升

◆ 提示与结果

> 🔊 **提示语（有删减）**

Please review and minimally edit the following sentences to ensure they are logically coherent, clear, and adhere to the conventions of academic writing in linguistics. Make minimal editing if you could.

"With the advent of large language models and AI assistants developed based on such models, we now have a brand new tool for collecting and investigating linguistic data, which will likely bring about new theoretical frameworks or new interpretations of existing frameworks, as well as a transformation, albeit probably gradual, in the practices of language learning and teaching."

适用模型：GPT-3.5、GPT-4

> ✅ **结果（GPT-3.5，有删减）**

The emergence of large language models and AI assistants, developed using such models, introduces a novel tool for the collection and analysis of linguistic data. This development is likely to give rise to new theoretical frameworks or fresh interpretations of existing ones, along with a transformation, albeit potentially gradual, in language learning and teaching practices.

> ✅ **结果（GPT-4，有删减）**

Your sentence is already well-structured and largely adheres to the conventions of academic writing in linguistics. However, for enhanced clarity and coherence, a few minimal adjustments can be made:

"With the advent of large language models and the development of AI assistants based on these models, we now possess a novel tool for collecting and analyzing linguistic data. This is likely to engender new theoretical frameworks or novel interpretations of existing ones, and to gradually transform language learning and teaching practices."

These changes improve the flow of the sentence and ensure precision in language, which is crucial in academic writing.

◆ 解读与反思

从结果来看，大语言模型能够按照提示语的要求识别写作文本中不自然的表达，也能对文本进行整体润色。但大模型修改的版本并不一定完全优于修改前的版本，使用者需仔细审阅结果。一方面，进一步精调提示语可以改善大模型的修改结果；另一方面，这也说明大模型使用者本身需要达到一定的语言水平。

此外，将学术写作文本上传大语言模型涉及用户数据的隐私和安全问题。使用者应充分了解各个大模型产品有关用户数据的使用条款，进行相关设置，尽量规避风险。

3.4.4 论文格式调整

3.4.4.1 英美英语转换

◆ 问题与目标

期刊对于论文语言可能存在一定的要求，大语言模型可帮助我们批量调整词汇选择、词汇拼写、标点符号使用等，在不同区域的英语变体间切换。本案例所用文本来自 Sinclair（2004：37）的专著。

🔊 提示语（有删减）

Could you revise this text to align with American English standards? Please focus on 1) correcting spelling, 2) adjusting the relative position of quotation marks and other punctuation marks, 3) serial comma usage, and 4) modifying the date format to the American style. Also, could you highlight in bold any words in the original text that have different meanings in British and American English, which should be double-checked for accuracy before final submission?

"The usual semantic gloss on *brook* as a verb is 'tolerate', and, as far as it goes, this is true. 'Tolerate' can replace *brook* in all the examples without disturbing their message. But *brook* is always negative; it expresses intolerance, not tolerance; the intolerance is of intrusive behaviour by another.
There is an inherent component of future modality in the expression, which shows that the possible intrusions are into plans or policies, and that the expression is a threat or warning. The displacement by report of the threat, and the frequent naming

of authority figures as subject of *brook*, complete the expression of a semantic prosody. There is another, rather elusive element of the prosody that we have not so far reconciled with the data. This phrase is emotionally charged with the commitment of the threatener to carry out the threat. Some of it may be in the words chosen as objects, especially where the negative comes in front of *brook – petulant isolation, challenge, protests, criticisms, defeat, contradiction* and *treachery*. In the other cases—the majority—the emotional charge is in the position of the negative, governing the noun rather than the verb."

适用模型：GPT-4

✅ 结果（有删减）

"The usual semantic gloss on **brook** as a verb is 'tolerate,' and, as far as it goes, this is true. 'Tolerate' can replace **brook** in all the examples without disturbing their message. But **brook** is always negative; it expresses intolerance, not tolerance; the intolerance is of intrusive **behavior** by another.

There is an inherent component of future modality in the expression, which shows that the possible intrusions are into plans or policies, and that the expression is a threat or warning. The displacement by report of the threat, and the frequent naming of authority figures as subject of **brook**, complete the expression of a semantic prosody. There is another, rather elusive element of the prosody that we have not so far reconciled with the data. This phrase is emotionally charged with the commitment of the threatener to carry out the threat. Some of it may be in the words chosen as objects, especially where the negative comes in front of *brook: petulant isolation, challenge, protests, criticisms, defeat, contradiction,* and *treachery*. In the other cases—the majority—the emotional charge is in the position of the negative, governing the noun rather than the verb."

Key changes made:
1. Spelling: "behaviour" to "behavior".
2. Punctuation: Adjusted the placement of quotation marks around 'tolerate', and added a serial comma in the list near the end of the text.
3. Date format: There were no dates in the original text to modify.
4. Words in bold (brook, behaviour) indicate terms that might have different meanings or usage in American English and should be double-checked.

◆ 解读与反思

　　大语言模型可以便捷地将文本语言惯例调整为英式或美式英语，为使用者节省大量时间精力。类似的格式、正字法等调整同样可以使用大模型自动完成。这类任务的提示语基本遵循同一原则，即充分列举需要调整的项目，以免大模型有所遗漏。

　　该案例还可用于外语教学，如基于大语言模型的阅读教学、写作教学。教师可以使用大语言模型，基于相同的内容生成拼写不同、标点不同的英式英语阅读写作材料和美式英语阅读写作材料，以引导学生对英式、美式英语之间的区别加以分辨。

3.4.4.2　引用前后对应

◆ 问题与目标

　　文献引用是学术写作的重要一环。有时引用文献数量较大，加之不同期刊对引文格式要求不一，我们可能会在自动生成书目后，解除书目与正文中引文的同步，直接对书目文本的格式进行编辑。此时可能出现引文与引用书目不对应的情况。在最终人工校对前，我们可以使用大语言模型对论文中的引用对应情况先行检查。

◆ 提示与结果

🔊 提示语（有删减）

Please review the attached draft of an academic journal article. Your task is to cross-check every reference in the main body of the article against the References at the end. Identify any discrepancies, including:
1. References in the text not listed in the References.
2. Reference items not mentioned in the text.
3. Inconsistencies in reference information, such as differing spellings of author names or publication years.

List all identified errors in a table with these columns: "Reference Label", "Reference Item", and "Description of Error".

适用模型：GPT-4

> 结果（有删减）

Reference Label	Reference Item	Description of Error
Juilland & Chang-Rodríguez (1964)	Juilland A. & Chang-Rodriguez E. (2021)	Incorrect publication year in the text. Should be 2021, not 1964.
Buchaan (1929)	Buchanan M. A. (1929)	Misspelled author's name in the text as "Buchaan".
Bou (1952)	Not Listed	Reference in the text not listed in the bibliography.
Chávez & Moreno (2016)	Chávez J. A. B. & Moreno D. A. H.(2018)	Incorrect publication year in the text. Should be 2018, not 2016.

◆ 解读与反思

大语言模型能够对论文中引用不匹配的多类情况进行识别，可辅助使用者的校对工作。但正如此前案例中提到的，使用大语言模型处理用户数据依然存在一定风险。如果文本内容敏感，使用者还需充分了解大模型产品的相关政策，谨慎衡量是否要将未发表的论文全文提供给在线模型。此外，大模型的校对也可能出错或遗漏，人工检查仍必不可少。

3.4.4.3 文献格式管理

◆ 问题与目标

本例旨在将参考文献格式转换为 GB/T 7714-2015 参考文献格式。

◆ 提示与结果

> 提示语（有删减）

Can you format the references following the style of GB/T 7714-2015? Please list all the references included.
【Example references】
BOULIANNE S. Revolution in the making? Social media effects across the globe [J]. Information, Communication & Society, 2019, 22(1): 39-54.
FIELD A. Discovering statistics using SPSS (3rd edition.) [M]. Los Angeles, CA: Sage Publications, 2009.

HILPERT M. Collostructional analysis: measuring associations between constructions and lexical elements [C]//GLYNN D, ROBINSON J. Corpus methods for semantics: quantitate studies in polysemy and synonymy. Amsterdam: John Benjamins, 2014: 391-404.
刘文宇，胡颖．基于文本挖掘的非传统文本批评话语研究[J]．天津外国语大学学报，2020（4）：29-41.
卫乃兴．词语学要义[M]．上海：上海外语教育出版社，2011.
王国成．计算社会科学：发展现状与前景展望[N]．中国社会科学报，2020-08-18（4）．
…
【References to be formatted】
(Please show the results in a table format, showing both the original reference and the formatted reference side by side. Highlight any changes in the formatted reference by making them bold.)
Boulton, A., & Cobb, T. 2017. Corpus use in language learning: A meta-analysis. Language Learning 2: 348–393.
Callies, M. (2019). Integrating corpus literacy into language teacher education. Learner Corpora and Language Teaching 92: 245-263.
…

适用模型：GPT-4

● 结果（有删减）

BOULTON A, COBB T. Corpus use in language learning: A meta-analysis [J]. Language Learning, 2017, 2: 348-393.
CALLIES M. Integrating corpus literacy into language teacher education [J]. Learner Corpora and Language Teaching, 2019, 92: 245-263.
…

◆ 解读与反思

在编写提示语时，也可给出 GB/T 7714-2015 参考文献格式的 5 至 10 个例子，以提高格式转换的准确率。

对于其他文献参考格式，只要在提示语中指明转换为 APA、MLA、Chicago、Harvard 等格式，大语言模型即可有效地进行参考文献的格式转换。

3.4.5 投稿修改完善

3.4.5.1 关键词语提取

◆ 问题与目标

论文关键词是研究的检索标签，需准确、充分地展示研究领域、对象和问题。如果在论文写作中对关键词的选择缺少把握，我们不妨使用大语言模型来启发思路或验证论文关键词。本案例展示如何借助大模型，基于论文摘要生成适宜的关键词。

◆ 提示与结果

🔊 提示语（有删减）

Based on the provided abstract of an academic journal article, could you suggest a group of keywords that best summarize the research?

"Traditionally, dictionaries are meaning-driven—that is, they list different senses (or supposed senses) of each word, but do not say much about the phraseology that distinguishes one sense from another. Grammars, on the other hand, are structure-driven: they attempt to describe all possible structures of a language, but say little about meaning, phraseology, or collocation. In both disciplines during the 20th century, the practice of inventing evidence rather than discovering it led to intermittent and unpredictable distortions of fact. Since 1987, attempts have been made in both lexicography (COBUILD) and syntactic theory (pattern grammar, construction grammar) to integrate meaning and phraseology. Corpora now provide empirical evidence on a large scale for lexicosyntactic description, but there is still a long way to go. Many cherished beliefs must be abandoned before a synthesis between empirical lexical analysis and grammatical theory can be achieved. In this paper, by empirical analysis of just one word (the noun *way*), we show how corpus evidence can be used to tackle the complexities of lexical and constructional meaning, providing new insights into the lexis-grammar interface."

适用模型：GPT-4

✅ 结果（有删减）

1. Lexicography: As the study focuses on dictionary practices and their evolution.

2. Syntax Theory: Mentioned in relation to grammatical structure and its relationship with meaning.
3. Corpus Linguistics: The use of corpora for empirical evidence is a central theme.
4. Lexico-Syntactic Description: A key focus of the research, integrating lexical analysis with grammatical theory.
5. Phraseology: The study appears to delve into how phraseology distinguishes meanings in language.
6. Pattern Grammar: Mentioned as a part of syntactic theory that integrates meaning and phraseology.
7. Construction Grammar: Also part of the syntactic theory discussion in the context of meaning and phraseology.
8. Empirical Evidence: Central to the methodological approach of the study.
9. Lexis-Grammar Interface: The intersection of lexicology and grammar is a main focus.
10. Semantic Analysis: Implied through the examination of lexical and constructional meaning.

◆ 解读与反思

该文章的实际关键词为 Corpus Pattern Analysis、empirical lexicography、construction grammar、pattern grammar、corpus evidence、collocation、local syntax、phraseology。

大模型生成的关键词与案例论文实际采用的关键词非常相似，说明大模型具备通过摘要提取关键词的能力。但大模型建议的关键词可能不够聚焦实际研究问题，也无法识别摘要中未直接体现的重要概念。因此，最终关键词的选择依然需要结合领域研究现状等多方面因素人工判断。

3.4.5.2 论文标题评估

◆ 问题与目标

论文标题往往决定读者对论文的第一印象，确保论文标题准确、简明地概括文章内容和视角至关重要。作为非母语者，我们可能无法准确感知拟定标题文本所传达的信息。此时，我们可以借助大语言模型对标题进行检查和打磨。

◆ 提示与结果

🔊 提示语（有删减）

Given the abstract of an academic journal article attached, please suggest improvements for its current title: "The Way to Analyze 'Way': A Case Study in Word-specific Local Grammar"?

"Traditionally, dictionaries are meaning-driven—that is, they list different senses (or supposed senses) of each word, but do not say much about the phraseology that distinguishes one sense from another. Grammars, on the other hand, are structure-driven: they attempt to describe all possible structures of a language, but say little about meaning, phraseology, or collocation. In both disciplines during the 20th century, the practice of inventing evidence rather than discovering it led to intermittent and unpredictable distortions of fact. Since 1987, attempts have been made in both lexicography (COBUILD) and syntactic theory (pattern grammar, construction grammar) to integrate meaning and phraseology. Corpora now provide empirical evidence on a large scale for lexicosyntactic description, but there is still a long way to go. Many cherished beliefs must be abandoned before a synthesis between empirical lexical analysis and grammatical theory can be achieved. In this paper, by empirical analysis of just one word (the noun *way*), we show how corpus evidence can be used to tackle the complexities of lexical and constructional meaning, providing new insights into the lexis-grammar interface."

适用模型：GPT-4

✓ 结果（有删减）

The current title, "The Way to Analyze 'Way': A Case Study in Word-specific Local Grammar", succinctly captures the essence of the paper's focus on a detailed analysis of the word "way" and its implications for understanding the intersection of lexicography and grammar. However, it could be improved to emphasize the innovative approach and broader implications of the study. Here are some suggestions:
1. "Redefining 'Way': A Corpus-driven Approach to Lexicosyntactic Analysis"
2. "Beyond Dictionaries and Grammars: A Corpus-based Study of 'Way' "
3. "Integrating Lexis and Grammar: Empirical Insights from the Analysis of 'Way' "
4. "Unpacking 'Way': New Perspectives on Lexical and Grammatical Interface"
5. "The Lexical Complexity of 'Way': A Corpus-based Exploration"
6. "Empirical Lexicography: The Case of 'Way' in Lexicosyntactic Context"

7. "Navigating 'Way': Corpus Evidence and the Lexis-Grammar Interface"
8. "Dissecting 'Way': A Case Study in Empirical Lexicosyntactic Theory"
9. "Way Forward: Corpus Analysis and the Synthesis of Lexical and Grammatical Theory"
10. "Exploring 'Way': A Groundbreaking Study in Corpus-based Lexicography"

These titles aim to highlight the innovative, empirical methodology used in the study and its contribution to understanding the complex relationship between lexis and grammar.

◆ 解读与反思

大语言模型可以依据摘要内容提供丰富的标题选择，使用者可根据实际需求进行判断。整体来说，大模型给出的标题更加中规中矩。在10个选项中，只有一个（Navigating 'Way': Corpus Evidence and the Lexis-Grammar Interface）采用了和论文实际标题类似的文字游戏，活用目标词way的意义构成标题。如果有明确的风格倾向，我们也可以在提示语中添加相关描述。同时，我们还可以进一步追问不同版本措辞的差异，比较、筛选出合适的标题。

3.4.5.3 目标期刊推荐

◆ 问题与目标

大语言模型可基于摘要内容快速筛选出与论文研究方向符合的备选期刊，既可以帮助我们验证当前论文是否符合已选定的目标期刊的要求，也可为投稿其他期刊提供参考。

◆ 提示与结果

🔊 提示语（有删减）

Based on the provided abstract of an academic journal article, what do you think are the most relevant journal that it should be submitted to? Please list the website of the recommended journals.

"Traditionally, dictionaries are meaning-driven—that is, they list different senses (or supposed senses) of each word, but do not say much about the phraseology that distinguishes one sense from another. Grammars, on the other hand, are structure-driven: they attempt to describe all possible structures of a language, but say little about meaning, phraseology, or collocation. In both disciplines during the 20th century, the practice of inventing evidence rather than discovering it led to intermittent and unpredictable distortions of fact. Since 1987, attempts have been made in both lexicography (COBUILD) and syntactic theory (pattern grammar, construction grammar) to integrate meaning and phraseology. Corpora now provide empirical evidence on a large scale for lexicosyntactic description, but there is still a long way to go. Many cherished beliefs must be abandoned before a synthesis between empirical lexical analysis and grammatical theory can be achieved. In this paper, by empirical analysis of just one word (the noun *way*), we show how corpus evidence can be used to tackle the complexities of lexical and constructional meaning, providing new insights into the lexis-grammar interface."

适用模型：GPT-4

✅ 结果（有删减）

The abstract you've provided indicates a focus on lexicography, syntax, and the integration of corpus evidence into linguistic analysis. Given this focus, journals specializing in linguistics, particularly those with an emphasis on lexicography, corpus linguistics, and grammatical theory, would be most relevant for submission. Here are a few recommendations:

1. *International Journal of Lexicography*: This journal focuses on all aspects of lexicography, making it a suitable outlet for research that integrates corpus evidence into lexical analysis.

2. *Corpus Linguistics and Linguistic Theory*: This journal is dedicated to corpus-based linguistic research, making it a good fit for papers exploring the empirical analysis of language.

3. *Journal of Linguistics*: This journal covers all aspects of linguistic research, including syntactic theory and grammar, which seems relevant given the focus on integrating lexical analysis and grammatical theory.

4. *Linguistics*: An interdisciplinary journal that covers various aspects of linguistic research, including both theoretical and empirical approaches.

These journals are reputable in the field of linguistics and provide an appropriate platform for research that seeks to bridge the gap between lexical analysis and grammatical theory through empirical methods.

◆ 解读与反思

大语言模型可根据摘要内容提供相关领域的期刊选择，辅助投稿过程。如果想要获得更有针对性的建议，我们可以要求大模型接入互联网信息，或从本地上传各个目标期刊对稿件的要求，同时在提示语中更详细地描述需考虑的因素，请大模型继续比较和筛选。

3.4.5.4 审稿意见模拟

◆ 问题与目标

国内外期刊审稿周期较长，作者往往难以得到及时的反馈或评价。面对这一现状，使用非母语进行学术写作的研究者尤为受阻。通过使用大语言模型，我们可以极为快捷地获取针对稿件的有用（且礼貌）的反馈（Liang et al. 2023），从而在正式投稿前进一步完善稿件。本案例使用的提示语改编自 Liang et al.（2023）的论文。

◆ 提示与结果

> 提示语（有删减）

Your task as an expert in the field is to draft a high-quality review for the provided article submitted to a top-tier linguistic journal. Your review should cover the following aspects: 1. Significance and novelty; 2. Potential reasons for acceptance; 3. Potential reasons for rejection (List multiple key reasons.) 4. Suggestions for improvement (List multiple key suggestions.) Be thoughtful and constructive. Write outlines only.（以 pdf 附件形式附上论文）

适用模型：GPT-4、Claude 2

> 结果（有删减）

The article titled "Modes of Meaning" presents an interesting exploration of the linguistic aspects of meaning in language. Here's an outline for a high-quality review for this article:
1. Significance and Novelty:
   - Significance: The article delves into how meanings are construed in linguistics, emphasizing the multi-faceted nature of meaning and its dispersion into various

modes, like phonetics, phonology, syntax, and semantics. This comprehensive approach is significant for understanding the complexity of language.
- Novelty: The innovative aspect lies in the detailed analysis of linguistic elements and their interplay in constructing meaning. The concept of "modes of meaning", which integrates various linguistic disciplines, is a novel approach in linguistic studies.

2. Potential Reasons for Acceptance:
   - Interdisciplinary Approach: The article's integration of different linguistic aspects (like phonetics, syntax, semantics) provides a holistic view of language analysis.
   - Depth of Analysis: The detailed examination of linguistic elements and their contribution to meaning is thorough and insightful.
   - Relevance: The topic is highly relevant in the field of linguistics, particularly in semantics and pragmatics.

3. Potential Reasons for Rejection:
   - Complexity and Accessibility: The article's depth might make it less accessible to a broader audience or those new to the field.
   - Empirical Evidence: The article might benefit from more empirical data or examples to support its theoretical framework.
   - Innovativeness: While the approach is comprehensive, it may not be considered groundbreaking in the field of linguistics, as it builds upon existing theories and models.

4. Suggestions for Improvement:
   - Empirical Examples: Incorporate more real-life language examples to illustrate the modes of meaning more concretely.
   - Clarity and Accessibility: Simplify some of the complex jargon for better accessibility to readers not deeply familiar with linguistic theories.
   - Comparative Analysis: Compare this approach to other models or theories in linguistics to highlight its uniqueness and contribution to the field.
   - Application: Demonstrate how this theoretical framework can be applied in practical linguistic analysis or other related fields like computational linguistics, language teaching, or psycho-linguistics.

Overall, the article presents a comprehensive and insightful perspective on language meaning, making a valuable contribution to linguistic scholarship. However, improvements in clarity, empirical grounding, and practical applications could enhance its impact and accessibility.

◆ **解读与反思**

　　大语言模型按照提示语要求，从多个维度对论文内容提出了具有参考价值的意见反馈。但正如 Liang et al.（2023）的观察，大模型更强调实证研究，部分评论较为宽泛，不够具体深入。这表明大模型模拟的审稿意见虽然可以作为作者投稿前的初步参考，但并不能取代实际流程中审稿人的意见。

第四章

结语

4.1　大语言模型外语教学与研究应用的挑战

在大模型高速发展的背景下，我们倡导大语言模型赋能的外语教学法和外语研究方法论的提质升级。这其中的指导原则应当是"人机协同"。人类的很多工作必然会为大模型取代，但这并不意味着人类的主体地位随之被取代。首先，大模型由人类创造并维护；其次，教学和研究问题的提出，始终由人发起；再次，教学材料和活动的设计，以及研究数据的采集和分析，都需要人的参与和监督；最后，大模型生成的教学内容和外语研究数据分析的结果，都需要人类专家验收裁定。人类节省的是时间，提高的是效率，但责任和价值丝毫没有减损。在可预见的将来，大模型必将持续引导外语学科创新。

今后一段时间的大模型研发可能会将责任安全、模态拓展、幻觉（hallucination）减除、模型解释等作为重点。概言之，目前我们看到的机遇多于挑战，我们应趋利避害、善用大模型。相关挑战和风险应予以重点关注，但这些是发展中的问题，一定能得到合理解决或管控，无须过度焦虑。日常我们听到不少对大模型的批评意见，这些应引起足够重视。随着大模型的发展，价值观念、伦理意识、信息安全已成为影响大众认知、学校教育，乃至治国理政的大事。负责任的人工智能（responsible AI）、可信赖的人工智能（trustworthy AI）、合乎伦理的人工智能（ethical AI）在学界和科技界已受到高度关注。相关立法、规范和技术标准已逐步推出或正在酝酿之中。如果没有安全可靠的人工智能管控机制，所有的技术成就都可能成为反噬人类的力量。我们相信政策制定者、技术公司、科研机构有智慧和能力筑牢责任安全的防线。可以预见，大模型的下一轮主要技术进步（如 GPT-5 和 Gemini Ultra，参见 Naddaf 2023）会以全面强化多模态理解和生成为特色。作为外语教学和研究人员，我们应当及早思考高性能多模态大模型给外语教学和研究带来的机遇。例如，我们会比以往任何时候都更有条件创设近似真实场景的体验式外语学习环境；我们可以跳出以书面语言分析为主的语言学研究窠臼，大张旗鼓地开展图文、图文音视频融合的多模态分析。基于文本的大模型建设，仍然会在扩大数据量方面继续发力。然而，比语料规模更关键的可能是生成内容的精准度和时效性。目前谈论较多的大模型幻觉也属于这方面的问题。随着算法的升级和优化，以及与搜索引擎的协同，大模型生成内容的质量必将得到保证。例如，最初版本的GPT-4模型所用训练语料截至2021年9月，而2023年9月27日更新的版本数据已更新至2023年4月，并增加了"借助必应浏览"（Browse with Bing）的功能。谷歌的Gemini大模型、百度的文心一言大模型也都结合了搜索引擎的事实查询功能。据此，静态数据的上限被打开，大模型也可以与最新数据进行交互。最后是大模型的可解释性问题。这是生成式人工智能领域的本体性议题，短期内不一定会有突破。对一般用户而言，大模型这个黑匣子为何如此强大无关紧要，然而，对于语言研究者来说，大模型为何能识别、标注、分类、解释语言现象，有着重要的理论价值。例如，赵冲（2023）尝试探究大模型在利用词向量技术进行词义辨析时，在哪些特定层级能够捕捉搭配信息，哪些层级可以存储句法信息，哪些层级可以承载语义语用信息。这为我们了解计算机科学家如何基于文本中的语符编码表征语义和人类知识提供了可能性。这样的大模型可解释性研究有助于我们更好地揭示语言规律，更好地将语言与人类知识、语言与文化联系起来。

大模型生成内容前后不一致的问题也值得关注。大模型每次都能生成与前次生成任务略有区别的内容。从语言产出的风格来看，这是有价值的，我们据此可以读到文风多样的文本。然而，对于推理分析和数据标注等任务，若前后两次操作结果有偏差，是不利于我们采信大模型、开展严肃科学研究的。当然，这一情况与大模型中的"温度系数"（temperature）这一设置参数有关。大模型和相关应用的开发者会专门设置一个体现结果波动的温度系数，使得不同用户采用相同提示语时可以得到差异化的反馈。我们在调用大模型API时，可以将温度系数的值设为0来解决这一问题，从而得到稳定、前后一致的输出内容。另外，大模型的版本迭代也会带来前后生成结果不一致的问题。不同的大模型系统输出的结果也会有所差别，因为不同的大模型在训练语料和算法实现方面有一定差异。

4.2 大语言模型外语教学与研究应用的机遇

大模型的发展还有一些突出特点。大模型的成功有赖于大数据和大算力，但这也为这项技术的推广和普及设置了技术壁垒。因此，相关技术公司也相继推出小模型，以及专门化和多版本大模型的技术开发路径。例如，GPT-4 中增添了"自定义指令"（custom instructions）、"自定义聊天机器人"（GPTs）功能。此外，大模型的原生应用也如雨后春笋般涌现。相信不久即会形成欣欣向荣的大模型生态系统，其中自然会有不少跟外语教学和外语研究相关的功能和应用。

随着大模型的普及，大模型生成的内容，又称"人工智能生成内容"（artificial intelligence generated content，简称 AIGC）也成为重要的语言研究对象。比如，北京外国语大学语料库语言学研究团队于 2023 年 8 月通过 GPT-3.5 生成库容为 200 万词的 aiTECCL 作文语料库，为我国中介语语料库研究提供了新型的近母语对照语言资源。另外，我们不得不正视的一个事实是，现实中的新闻话语、学术话语、翻译文本都有可能是由大模型完成，而非人工写成的。当前混杂"专业人士生成内容"（professional generated content，简称 PGC）、"用户生成内容"（user generated content，简称 UGC）和人工智能生成内容的语言生活可能会成为常态，甚至会出现人工智能生成内容占主导的可能性。这些应引起足够重视。

目前大模型的主要使用方式是用户通过网页界面或手机应用与之互动。这样的操作流程已完全能满足通常的语言教学和简单的研究任务。然而，相关大模型对输入和上传提示语内容的容量均有上限要求。在外语研究中，我们时常需要对大规模数据和文本进行分析。比如，若要对 100 万词的英文文本进行情感分析，或从几百页的专业英文资料中抽取英文术语并提供对照翻译，逐段将文本粘贴到输入框并不现实。为此，相关大模型都提供了应用程序编程接口，用户只需进行少量编程并适当付费，即可大批量地生成内容或分析数据。

大模型的普及必将深刻影响每个人。在科技公司的人工智能军备竞赛之下，新的大模型不断推出，模型版本不断更新，智能体（AI agent）得到开发，免费可用的大模型资源比比皆是。在对互联网资源的利用上，我们已经从"搜索引擎时代"进入"大模型时代"。在日常办公、教学和科研中，我们已经从文字处理阶段进入以大模型为核心的"人机协同"新发展阶段。在数字素养方面，"问商"（prompting quotient）已有取代"搜商"（search quotient）之势，成为决定人们获取新知、不断进步的关键能力。

图 4.1　问商构成要素示意图

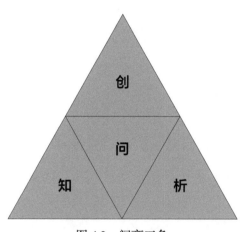

图 4.2　问商三角

其中，"知"和"析"是前提和基础，"问"是核心，"创"是最高目标。可见，问商不仅是电脑操作技术，它包括"知""析""问""创"等子能力。大模型及其应用将是新的历史阶段应有的学习和生活方式。我们不应放弃基础知识和常识的积累，这是一切新知的起点，更深厚的知识基础意味着我们处于更高的提问和创新平台之上。我们应以提示语探索大模型，高效率地解答疑问，从而更好地实现人类价值。

在上述背景下，外语教学和外语研究是否会发生本体性的变化，目前还未可知。有不少人对整个学科的后续发展产生忧虑，这主要是源于大模型在语言表达和交流能力方面的超强表现。焦虑情绪是对迅猛发展的人工智能技术的正常反应。如今，大模型已是"屋中大象"（elephant in the room），我们无法忽视，只有合理应用、正确引导、为我所用才是正途。我们需要进一步思考和探索的是，如何在充分了解大模型效能的基础上，构建出以服务人类为根本目标、以人类为主导的人机协同外语学习和外语研究新模式。在操作层面，熟悉大模型、利用大模型应当先从编写提示语入手。本书正是为外国语言学及应用语言学学科写成的一本提示工程入门书籍。

参考文献

Austin, J. 1962. *How to Do Things with Words*. Oxford: Clarendon Press.

Blei, D., A. Ng & M. Jordan. 2003. Latent Dirichlet allocation. *Journal of Machine Learning Research* 3: 993-1022.

Corder, S. 1967. The significance of learners' errors. *International Review of Applied Linguistics in Language Teaching 5* (4): 161-170.

Divjak, D. & N. Fieller. 2014. Cluster analysis: Finding structure in linguistic data. In D. Glynn & J. Robinson (eds.). *Corpus Methods for Semantics: Quantitative Studies in Polysemy and Synonymy*. Amsterdam: John Benjamins. 405-441.

Dörnyei, Z. 2005. *The Psychology of the Language Learner: Individual Differences in Second Language Acquisition*. London: Lawrence Erlbaum Associates.

Dörnyei, Z. 2009. The L2 motivational self system. In Z. Dörnyei & E. Ushioda (eds.). *Motivation, Language Identity and the L2 Self*. Bristol: Multilingual Matters. 9-42.

Floridi, L. & M. Chiriatti. 2020. GPT-3: Its nature, scope, limits, and consequences. *Minds & Machines 30* (4): 681-694.

Glynn, D. 2014. Correspondence analysis: Exploring data and identifying patterns. In D. Glynn & J. Robinson (eds.). *Corpus Methods for Semantics: Quantitative Studies in Polysemy and Synonymy*. Amsterdam: John Benjamins. 443-485.

Granger, S. 1998. The computer learner corpus: A versatile new source of data for SLA research. In S. Granger (ed.). *Learner English on Computer*. New York: Routledge. 3-18.

Halliday, M. & R. Hasan. 1976. *Cohesion in English*. London: Longman.

Jones, C. 2015. *The Dig*. Minneapolis: Coffee House Press.

Labov, W. & J. Waletzky. 1967. Narrative analysis: Oral versions of personal experience. In J. Helm (ed.). *Essays on the Verbal and Visual Arts*. Seattle: University of Washington Press. 12-44.

Lakoff, G. & M. Johnson. 1980. *Metaphors We Live By*. Chicago: University of Chicago Press.

Liang, W., Y. Zhang, H. Cao, *et al.* 2023. Can large language models provide useful feedback on research papers? A large-scale empirical analysis. arXiv preprint arXiv:2310.01783.

Lin, P. 2023. ChatGPT: Friend or foe (to corpus linguists)? *Applied Corpus Linguistics 3* (3): 1-5.

Martin, J. 2000. Beyond exchange: Appraisal systems in English. In S. Hunston & G. Thompson (eds.). *Evaluation in Text: Authorial Stance and the Construction of Discourse*. Oxford: Oxford University Press. 142-175.

Martin, J. & P. White. 2005. *The Language of Evaluation: Appraisal in English*. Basingstoke: Palgrave Macmillan.

Moran, P. 2001. *Teaching Culture: Perspectives in Practice*. Boston: Heinle & Heinle.

Naddaf, M. 2023. The science events to watch for in 2024. *Nature* 18 December.

Sinclair, J. 2004. *Trust the Text: Language, Corpus and Discourse*. New York: Routledge.

Sinclair, J. & M. Coulthard. 1975. *Towards an Analysis of Discourse: The English Used by Teachers and Pupils*. Oxford: Oxford University Press.

Swales, J. 1990. *Genre Analysis: English in Academic and Research Settings*. Cambridge: Cambridge University Press.

Toulmin, S. 1958. *The Uses of Argument*. Cambridge: Cambridge University Press.

Xu, J. & H. Kang. 2022. Salience-simplification strategy for markedness of causal subordinators: "Because" and "since" in argumentative essays. *Lingua* 272: 103256.

Xu, J. & J. Li. 2021. A syntactic complexity analysis of translational English across genres. *Across Languages and Cultures* 22 (2): 214-232.

鲍刚，1996，译前准备"术语强记"的方法论，《北京第二外国语学院学报》（3）：42-45。

陈向明，2000，《质的研究方法与社会科学研究》。北京：教育科学出版社。

方梦之，2018，中外翻译策略类聚——直译、意译、零翻译三元策略框架图，《上海翻译》（1）：1-5。

桂诗春，2004，以语料库为基础的中国学习者英语失误分析的认知模型，《现代外语》（2）：129-139。

桂诗春、杨惠中，2003，《中国学习者英语语料库》。上海：上海外语教育出版社。

何安平、许家金、张春青，2020，《语料库辅助中学英语教学案例选编》。北京：外语教学与研究出版社。

姜琳、涂孟玮，2016，读后续写对二语词汇学习的作用研究，《现代外语》（6）：819-829。

蒋平，2004，零形回指现象考察，《汉语学习》（3）：23-28。

邱琳，2017，"产出导向法"语言促成环节过程化设计研究，《现代外语》（3）：386-396。

邱琳，2020，"产出导向法"促成环节设计标准例析，《外语教育研究前沿》（2）：12-19。

孙曙光，2019，"师生合作评价"的辩证研究，《现代外语》（3）：419-430。

王初明，2016，以"续"促学，《现代外语》（6）：784-793。

王初明，2017，从"以写促学"到"以续促学"，《外语教学与研究》（4）：547-556。

王初明，2018，续译——提高翻译水平的有效方法，《中国翻译》（2）：36-39。

王初明，2019，运用续作应当注意什么？《外语与外语教学》（3）：1-7。

王启、王初明，2019，以续促学英语关系从句，《外语教学理论与实践》（3）：1-5。

文秋芳，2015，构建"产出导向法"理论体系，《外语教学与研究》（4）：547-558。

文秋芳，2016，"师生合作评价"："产出导向法"创设的新评价形式，《外语界》（5）：37-43。

文秋芳、孙曙光，2020，"产出导向法"驱动场景设计要素例析，《外语教育研究前沿》（2）：4-11。

辛声，2017，读后续写任务条件对二语语法结构习得的影响，《现代外语》（4）：507-517。

熊淑慧，2018，议论文对比续写的协同效应研究，《解放军外国语学院学报》（5）：85-92。

许家金，2019，《语料库与话语研究》。北京：外语教学与研究出版社。

许家金等，2023，《语料库研究方法》。北京：外语教学与研究出版社。

许琪，2016，读后续译的协同效应及促学效果，《现代外语》（6）：830-841。

杨华，2018，读后续写对中高级水平外语学习者写作修辞的学习效应研究，《外语教学与研究》（4）：596-607。

杨士焯，2006，《英汉翻译教程》。北京：北京大学出版社。

张培基、喻云根，1980，关于《英汉翻译教程》的编写，《教学研究》（2）：52-54。

张文娟，2017，"产出导向法"对大学英语写作影响的实验研究，《现代外语》（3）：377-385。

张秀芹、王迎丽，2020，读后续说任务中语言水平对学习者输出及协同效果的影响，《解放军外国语学院学报》（1）：9-16。

赵冲，2023，面向学习词典义项识别的语境嵌入研究。博士学位论文。北京外国语大学。

朱奕瑾、饶高琦，2023，基于 ChatGPT 的生成式共同价值标准例句库建设，《云南师范大学学报（对外汉语教学与研究版）》（3）：71-80。

后记

本书责任编辑发来清样时,一并附上了封面设计方案,征询我的看法。图案简明大方,然而终归觉得设计方案主题不显,泯然于众。出于对"有意味形式"(significant form,Bell 1913:8)的追寻,我再度思考《大语言模型的外语教学与研究应用》一书的主旨。

本书旨在倡导利用大语言模型"预测下一个词"的特性,通过编写提示语,促进语言教学和研究的发展。因此,我希望本书封面可以彰显以下意涵。

语言如行棋。在学术史上,很多学者曾以行棋类比语言运作。例如,学界常以这句话概括索绪尔的结构主义思想:"The respective value of the pieces depends on their position on the chessboard just as each linguistic term derives its value from its opposition to all the other terms"(Saussure 1916/1959:88;GPT-4译文:棋子在棋盘上的各自价值,取决于它们的位置,正如每个语言符号的价值,皆源自与其他词的对立)。维特根斯坦更热衷于用象棋来阐释语言规律,并提出了"语言游戏观"。他指出:"Words and chess pieces are analogous; knowing how to use a word is like knowing how to move a chess piece"(Ambrose 2001:3;GPT-4译文:文词与棋子相映;知词之用,犹如知棋之行也)。鉴于此,我采用国际象棋棋盘这一具象符号作为封面图案,同时棋盘中嵌入"纵横字谜"。将国际象棋与字谜游戏相结合,旨在凸显语言的结构性、系统性及互动性。

智能似处子。人工智能发展史上,很多重要的智能创新都以战胜人类棋手为标志(Pickover 2019)。比如,阿尔法围棋(AlphaGo)战胜李世石和柯洁,深蓝(Deep Blue)战胜卡斯帕罗夫;再如,1959年,亚瑟·塞缪尔提出"机器学习"这一概念时,也以行棋决策来阐述其算法。可见,人工智能与棋类游戏关系密切。

模型即分布。 基于大语言模型的人工智能有其语言学机理，即 Harris（1954）提出的"分布假说"（distributional hypothesis）。大语言模型依托海量数据和强大算力，通过数学建模挖掘语言分布规律，深刻描绘语言知识与世界知识。分布规律亦即语境或上下文。封面图案中的字谜设计便是运用此原理，借由上下文推敲确切词汇。

提示乃问商。 在封面图案的纵横字谜部分，我将本书的五大核心主题，即大语言模型（LLM）、语言（language）、教学（teaching）、研究（research）和提示语（prompt），分别编号 1 至 5，以展示大语言模型如何基于"语符编号"（token ID）预测下一词。其中，大语言模型作为方法引领，居于首位；语言是其数据基础；教学与研究构成本书两大关键主题；提示语是模型应用的抓手，收官压轴。在本书的结语中我也曾特别强调，进入人工智能发展新阶段，"问商"和"以问促创"在教育与研究中的作用将会十分突出。

语料可成事。 在黑白两色之外，本书以橙色作为主要色彩基调。橙色为我所在的北京外国语大学语料库语言学研究队伍的团队色，称为"语料橙"，借此寓意"语料可成事"。

基于上述思考，我编写了提示语[1]，利用 GPT-4 生成了若干封面图案。从中优选了部分图样作为设计草案，交由专业美术编辑设计成型。这一过程中，我根据学科知识和图书主题提炼需求，并采用大语言模型以文生图，精准对接美术编辑，从而实现"图书作者＋人工智能＋美术编辑"三方协作的封面设计模式。这是"人机协同大语言模型方法论"的一次生动实践。

---

1 封面初稿设计提示语：请帮我设计一个大语言模型应用于语言学的学术图书封面。主体图案是一个国际象棋棋盘，棋盘上是 crossword puzzle 的样式。主体颜色为橙色。请采用极简风格。

后记参考文献

Ambrose, A. (ed.). 2001. *Wittgenstein's Lectures: Cambridge, 1932–1935: From the Notes of Alice Ambrose and Margaret Macdonald*. New York: Prometheus Books.

Bell, C. 1913. *Art*. New York: Frederick A. Stokes Company.

Harris, Z. 1954. Distributional structure. *Word 10* (2-3): 146-162.

Pickover, C. 2019. *Artificial Intelligence: An Illustrated History from Medieval Robots to Neural Networks*. New York: Sterling.

Samuel, A. 1959. Some studies in machine learning using the game of checkers. *IBM Journal of Research and Development 3* (3): 210-229.

Saussure, F. 1916/1959. *Course in General Linguistics*, ed. C. Bally & A. Sechehaye, trans. W. Baskin. New York: The Philosophical Library.